INSIDE INFORMATION
Computers in Fiction

INSIDE INFORMATION
Computers in Fiction

ABBE MOWSHOWITZ
University of British Columbia

ADDISON-WESLEY PUBLISHING COMPANY
Reading, Massachusetts · Menlo Park, California
London · Amsterdam · Don Mills, Ontario · Sydney

Copyright © 1977 by Addison-Wesley Publishing Company, Inc. Philippines copyright 1977 by Addison-Wesley Publishing Company, Inc. All rights reserved. No part of this publication may be reproduced, stored in a retrieval system, or transmitted, in any form or by any means, electronic, mechanical, photocopying, recording, or otherwise, without the prior written permission of the publisher. Printed in the United States of America. Published simultaneously in Canada. Library of Congress Catalog Card No. 76-54429.

Reproduced by Addison-Wesley from camera-ready copy prepared by the author.

This book is in the
ADDISON-WESLEY SERIES IN COMPUTER SCIENCE
Michael A. Harrison
Consulting Editor

Acknowledgments

Grateful acknowledgment is made to the following for permission to quote from copyright material:

Brian Aldiss, "Full Sun." Copyright © 1967 by Damon Knight, first appeared in *Orbit 2*. Reprinted by permission of the author and A.P. Watt & Son.
Isaac Asimov, "Franchise." Copyright © 1955 by Quinn Publishing Co., Inc. Reprinted by permission of the author.
Isaac Asimov, "The Feeling of Power." Copyright © 1957 by Quinn Publishing Co., Inc. Reprinted by permission of the author.

ACKNOWLEDGMENTS v

Raymond E. Banks, "Walter Perkins Is Here!" Copyright © 1970 by Raymond E. Banks. From the *Future is Now* edited by William F. Nolan. Reprinted by permission of William F. Nolan.

John Barth. Selection from *Giles Boat-Boy*, copyright © 1966 by John Barth. Reprinted by permission of Doubleday & Company, Inc., New York, and Martin Secker & Warburg Limited, London.

Donald Barthelme, "Report." Reprinted with the permission of Farrar, Straus & Giroux, Inc. from *Unspeakable Practices, Unnatural Acts* by Donald Barthelme, Copyright © 1967, 1968 by Donald Barthelme; "Report" originally appeared in *The New Yorker*. Also reprinted by permission of Jonathan Cape Ltd, London.

Gregory Benford, "Nobody Lives on Burton Street," *Amazing Stories*, 1970. Copyright © 1970 by Ultimate Publishing Co. Reprinted by permission of the author and the author's agent, Richard Curtis.

Richard Brautigan, "All Watched Over by Machines of Loving Grace." Excerpted from *The Pill Versus The Springhill Mine Disaster* by Richard Brautigan. Copyright © 1968 by Richard Brautigan. Reprinted by permission of DELACORTE PRESS/ SEYMOUR LAWRENCE, and Jonathan Cape Ltd, London.

Frederic Brown, "Answer." Copyright © 1954 by Frederic Brown. Reprinted by permission of the author and his agents, Scott Meredith Literary Agency, Inc., 845 Third Avenue, New York, New York 10022.

John Brunner. Selection from *The Jagged Orbit* by John Brunner, copyright © 1969 by Brunner Fact & Fiction Ltd. Reprinted by permission of the author and his agents, Scott Meredith Literary Agency, Inc., 845 Third Avenue, New York, New York 10022.

Eugene Burdick. Selection from *The 480* by Eugene Burdick. Copyright © 1964 by Kamima, Inc. Used with permission of McGraw-Hill Book Company.

Arthur C. Clarke, "The Nine Billion Names of God." Copyright © 1953 by Ballantine Books, Inc. Reprinted by permission of the author and the author's agents, Scott Meredith Literary Agency, Inc., 845 Third Avenue, New York, New York 10022.

Arthur C. Clarke. Selections from *The City and the Stars*, copyright © 1953, 1956, by Arthur C. Clarke. Reprinted by permission of Harcourt Brace Jovanovich, Inc., and Victor Gollancz Ltd.

David G. Compton. Selection from *The Steel Crocodile* copyright © 1970 by D.G. Compton, published by Ace Books, and by Hodder & Stoughton (as *The Electric Crocodile*). Reprinted by permission of the author.

Lester del Rey, "Instinct." Copyright © 1951 by Street & Smith Publications, Inc. Reprinted by permission of the author and the author's agents, Scott Meredith Literary Agency, Inc., 845 Third Avenue, New York, New York 10022.

Samuel R. Delany. Selection from *Nova*, copyright © 1968 by Samuel R. Delany. Reprinted by permission of Doubleday & Co., Inc. Also used by permission of the author and Henry Morrison Inc., his agents.

Philip K. Dick. Selection from "Top Stand-By Job." Copyright © 1963 by Ziff-Davis Publishing Co., Inc. Reprinted by permission of the author and the author's agents, Scott Meredith Literary Agency, Inc., 845 Third Avenue, New York, New York 10022.

Gardner R. Dozois, "Machines of Loving Grace." Copyright © 1972 by Damon Knight, first appeared in *Orbit 11*. Reprinted by permission of the author and the author's agent, Virginia Kidd.

Hal Draper, "Ms Fnd in a Lbry." Copyright © 1961 by Mercury Press, Inc. Reprinted by permission of the author from *The Magazine of Fantasy and Science Fiction*.

Ron Goulart. "Hardcastle" is reprinted by permission of Charles Scribner's Sons from *What's Become of Screwloose* by Ron Goulart. Copyright © 1971 Ron Goulart. Also reprinted by permission of Sidgwick & Jackson Ltd, London.

Robert A. Heinlein. Selection from *The Moon is a Harsh Mistress*. Reprinted by permission of G.P. Putnam's Sons from *The Moon is a Harsh Mistress* by Robert A. Heinlein. Copyright © 1966 by Robert Heinlein. Also reprinted by permission of John Farquharson Ltd, London.

Christopher Hodder-Williams. Selection from *Fistful of Digits* copyright © 1968 by Christopher Hodder-Williams, first published by Hodder & Stoughton Ltd. Reprinted by permission of the author.

Olof Johannesson. Selection from *The Tale of the Big Computer*. Reprinted by permission of Coward, McCann & Geoghegan, Inc. from *The Tale of the Big Computer* by Olof Johannesson. Copyright © 1968 by Victor Gollancz Ltd. Also reprinted by permission of Victor Gollancz Ltd, London, and the author.

ACKNOWLEDGMENTS

Bruce Kawin, "FORM 5640A: Report of a Malfunction." Reprinted by permission of North American Publishing Co. from *Data Processing Magazine*.

R.A. Lafferty, "What's the Name of That Town." Copyright © 1964 by Galaxy Publishing Corp.; reprinted by permission of the author and his agent, Virginia Kidd.

Felicia Lamport, "A Sigh for Cybernetics." Copyright 1960 by *Harper's Magazine*. Reprinted by permission of the author.

Ira Levin. Selection from *This Perfect Day*, by Ira Levin. Copyright © 1970 by Ira Levin. Reprinted by permission of Random House, Inc., and Harold Ober Associates Incorporated.

Elmer L. Rice. Selection from *The Adding Machine* by Elmer L. Rice. Copyright, 1922, 1929, by Elmer L. Rice. Copyright, 1923, by Doubleday, Page & Company. Copyright, 1949, (In Renewal), by Elmer Rice. Copyright, 1956 (In Renewal), by Elmer Rice. All rights reserved. Reprinted by permission of Samuel French, Inc.

Louis B. Salomon, "Univac to Univac (*sotto voce*)." Copyright 1958 by Harper's Magazine. Reprinted by permission of the author.

Robert Sheckley. Excerpts from *Journey Beyond Tomorrow* by Robert Sheckley. Copyright © 1962 by Robert Sheckley. Reprinted with the permission of Dell Publishing Co., Inc., and The Sterling Lord Agency, Inc.

Alan Simpson, "The Twenty-Third Psalm— Modern Style." Reprinted by permission of *Washington University Magazine*.

Henry Slesar, "Examination Day." Originally appeared in *Playboy Magazine*; copyright © 1958 by Playboy. Reprinted by permission of the author.

Kurt Vonnegut, Jr. Chapter XI excerpted from *Player Piano* by Kurt Vonnegut, Jr. Copyright © 1952 by Kurt Vonnegut, Jr. Reprinted by permission of DELACORTE PRESS/SEYMOUR LAWRENCE. Also reprinted by permission of Kurt Vonnegut, Jr. and his Attorney, Donald C. Farber, 800 Third Avenue, New York, New York 10022.

In memory of my sister

Eleanor Mowshowitz Friend

Preface

Extensions of man's hegemony over nature are rarely conservative. Innovation, even when dictated by social necessity, creates disturbances which demand new modes of accommodation. Unfortunately, there is no better method than trial-and-error for achieving stable equilibria. This condition of uncertain experimentation characterizes the present state of computer utilization. Coping with the prodigious advance of the technology and the diffusion of computer applications throughout contemporary society is likely to exercise our skill and imagination for some time to come. Automation is altering the nature of work. Computer-based information systems promise to yield radical changes in education and health-care. Corporate and governmental administrative practices are undergoing major modifications in response to new methods of information processing. The computer has not yet invaded the home, but if computer-communications networks are developed on a large scale, the remote terminal will become as common as the ubiquitous television receiver. Beyond these concrete changes, both actual and potential, we may anticipate fundamental shifts in social evolution.

Computer applications exert an influence on the structure of decision-making and on the individual's relationship to the political process. The new technology is widely regarded as a timely response to the problems of managing mass society. Computers make it possible for government agencies and private organizations to process the mountains of information generated by an ever increasing volume of transactions. Considerations of efficiency seem to warrant large-scale centralized operations, and computers provide the means for coordination and control. As large organizations assume a greater share of the responsibility for social services, they also acquire greater power and authority, thus altering the balance between individual and community. The erosion of personal privacy, abetted by computerized databanks containing information on millions of people, is but one manifestation of the shifting boundary between public and private. Perhaps of greater moment in the long-run is the computer's influence on attitudes and values. The growing remoteness of the centers of public policy-making from the average citizen is matched by an increasing estrangement of the individual from traditional sources of self-affirmation. The possibility of being replaced by a machine undermines self-esteem; and there is little consolation in knowing that the prohibitive cost of automation is the only guarantee against the mechanization of one's function in society. Human identity is under siege. Whether or not "thinking machines" capable of challenging man's ascendency will ever materialize, there do exist computer programs which exhibit behavior normally believed to require intelligence in humans. The shrinking sphere of human uniqueness is a curious by-product of technological prowess. As our artifacts become more sophisticated, we become more inclined to draw comparisons between men and machines— success creates the conditions of its own justification.

    The unresolved issues associated with the computer reveal the depth of our uncertainty over how to make appropriate use of this extraordinary invention. We are on the threshold of major, new social commitments, and are baffled by imponderables which obscure our vision of the future. Experimental computer networks have demonstrated the technical feasibility of developing information utilities. One projected application which has aroused widespread interest and concern is the implementation of Electronic Funds Transfer Systems. The controversy surrounding this application illustrates the deficiencies in our understanding of the social implications of computer technology. Not only are we unable to dispel misgivings over potential threats to privacy and political integrity, but we cannot even resolve the more pragmatic issues of cost-effectiveness. Other applications pose yet greater problems. The use of computer utilities to promote citizen participation in the political process might require fundamental changes in the structure of social institutions. There is of course no formula for anticipating changes of this magnitude and complexity—works of fiction are often as insightful as scientific studies. Indeed, in the search for understanding, an active imagination may be a more valuable asset than a well-defined methodology.

This study-anthology is offered as a guide to imaginative literature on the social impact of computing. Fictional treatments of the computer are used as a vehicle for illuminating the discussion of social issues. Naturally, one cannot expect to find definitive answers— the intellectual contribution of computer tales lies in their inventiveness and in their openness to unconventional alternatives. Technical problems associated with computer applications are less important than the human dimensions of the issues. Ascertaining desirable uses of the technology depends more on understanding social conditions and priorities than it does on familiarity with the latest refinements in hardware and software engineering. A secondary emphasis of the work directs attention to contemporary attitudes towards computers. Attitudes constitute a sensitive barometer of changing social arrangements, and also point to value-conflicts produced or sharpened by technological innovation.

Lest the reader form erroneous expectations, two disclaimers must be aired. First, this is not a work of literary criticism. Apart from the author's judgment of what selections to include, aesthetic considerations are entirely subservient to social analysis. Secondly, the book is not intended as an introduction to computing— there is only a weak insistence on fidelity to technological fact. The literature is exploited for its insights into social and human problems linked to the use of computers. Fanciful departures from the present state of the art are sometimes warranted as dramatic expedients. By stretching the plausible through exaggeration or deliberate distortion we are able to see some issues in better perspective.

With few exceptions the stories discussed in this book were written after World War II. Computer-related fiction began in earnest in the early 1950's when the first production model computers appeared on the scene. Most of the literature would undoubtedly be classified as science fiction or science fantasy, but the definition of the literary genre to which the stories belong is quite peripheral to present concerns. Since computers play a casual role in a large segment of contemporary science fiction, it is not too useful to insist upon a rigid demarcation of the "computer tale." Although the distinctions between robots, androids, and computers are not always observed in science fiction, these concepts do provide a rough typology. In his introduction to *Science Fiction Thinking Machines* Conklin characterizes a robot as an essentially mechanical creature whose behavioral repertoire is circumscribed by built-in design limitations. Robots usually appear as servants, although they occasionally disobey "The Three Laws of Robotics" (see Asimov's *I, Robot*) formulated by Isaac Asimov in the early 1940's. Again following Conklin we may describe androids (or humanoids) as living beings "created partly or wholly through processes other than human birth." Androids typically exhibit human characteristics, and are often human in appearance. Since robots and androids often feature computerized control mechanisms or "brains," stories dealing with them must be included among the computer-related tales.

The most readily identifiable type of computer fiction deals with machines very much like those in current use. Extrapolations based on projected computer developments also present little difficulty. Problems arise when computers are endowed with capabilities that depart radically from present-day technological reality. Here is where the science fiction computer merges with the robot and the android. Robots, for example, may be viewed as mobile computers. In the realm of fantasy, virtually anything is possible and most classification schemes break down. Hybrid entities, of which the cyborg is a special case, combine organic and electronic or mechanical parts. Creatures which are part human brain and part computer force one to make arbitrary distinctions. Happily, this is not too disturbing since our main interest is in fictional accounts of the effects of mechanization and automation on social organization and human behavior.

The vision of man's future that emerges in the literature is not a comforting one. It is riddled with anxiety over the diminishing role of human beings in an increasingly technological world. The sense of malaise which permeates computer stories suggests an attitude of morbid fatalism. To some writers

PREFACE   xv

the waning fortunes of mankind signify nothing more than a passing phase in the evolution of cosmic mind; others are less sanguine in their expectations. In any case one finds a near universal belief in the inevitability of continued development of ever more powerful machines. Contemporary fiction foresees the creation of intelligent computers which will rival and perhaps replace man as the dominant species on earth. There are of course many steps on the road to oblivion, and not all stories are set in the remote future; but even the intermediate stages are imbued with a fatalistic acceptance of humanity's ultimate demise.

Fictional computers, like their real-life models, have been introduced into every corner of human affairs. The most noticeable effect is the displacement of human beings from economic and social roles that have been man's burden and prerogative since time immemorial. Automation is not confined to the production of goods— it extends to government decision-making, medical care, education, and even to the creative arts. No area of human endeavor is entirely immune to computerization. There are machines to govern, to heal the sick, to provide for material needs, to offer spiritual guidance, to care and to love, and to minister to creative impulses. Just as the aged are superfluous in industrial society, so are human beings in the unfolding electronic paradise. Since worth is measured by productivity, one finds little to recommend human survival.

Although it does give implicit recognition to opportunties and accomplishments, computer fiction dwells on the more problematic aspects of man-machine interaction. Computer technology serves as a microcosm of technology as a whole, and it is heir to all of the misgivings that have been expressed in anti-utopian writing. Information-processing systems are seen to intensify the dehumanizing influences of industrialization. In particular, computer applications are shown to be insensitive to human needs and to ignore human abilities— systems are designed to increase productivity and efficiency, not to improve the quality of life. Because of its easily identifiable role in corporate and government bureaucracy, the computer bears the brunt of the anger and frustration endemic to life in the mass society. The kinds of problems highlighted in fiction are characteristic of peoples' experience with computers, i.e., being on the receiving end of computerized billing, payroll, magazine subscription, direct mail advertising, etc. Most of the time these applications are unobtrusive, but when errors occur in badly designed systems the impersonality of the machine becomes a major issue.

Much of the justification for negative appraisals of the computer is reminiscent of pre-scientific modes of thought. To the pre-modern mind the primordial forces of nature were manifest not in the regularities of natural phenomena but in miraculous occurrences. The focus on computer system errors and malfunctions leads to a distorted picture of reality in which untoward events define normal practice. Although it may have dramatic impact, this approach to depicting human problems in computerized society undermines the validity of the observations. If computer-based information systems do indeed act as an impediment to communication, it is not primarily for reasons of technical design limitations. The ludicrous inflexibility often represented in stories about computers is embedded in the structure of the social institutions which make use of the technology.

In contemporary fiction the remorseless extension of technology is virtually synonymous with the growing superfluousness of human beings. Succeeding generations of machines have reduced the need for human strength, motor skills, and now intellectual abilities. The sphere of individual moral choice has contracted along with the shrinkage in the arena for human action. This observation surfaces in the form of a quest for identity— a sometimes desperate struggle to resolve the ambiguities of man's role in modern society and his ambivalence towards machines. By putting us out of work and usurping our creative leisure, machines rob us of the most vital supports of personality. The compulsion to displace human labor wherever possible is entirely blind to the centrality of work in maintaining a stable ego. Unfortunately, no viable alternatives have yet been found.

Modern technology also impinges on human identity in a direct way. The possibility of intelligent computers forces us to re-examine our beliefs in the uniqueness of human attributes. It becomes more difficult to accept the idea of some special dispensation for the human race. Considerations of this kind constitute a major preoccupation of computer-related fiction. One meets computers in every stage of "human" development, from the child's differentiation of self from others, through the adolescent's exploration of personal identity, to the mature consciousness of adulthood, and beyond. Yet in spite of their remarkable abilities most of these machines are drawn as human beings *manqué*. Science fiction writers are mesmerized by the concept of machine consciousness, and they are unable to allow their creations sufficient freedom to achieve an independent sense of self. A related form of this failure of nerve is evident in stories of human regeneration by machine. This reflects a psychological paradox— the ascent of machines appears inexorable, yet it is inconceivable that man should vanish or become a subsidiary creature.

Fear of the unknown rivals the challenge to identity as a pervasive theme in the literature. Recent history shows all too painfully that even our most well-intentioned efforts may have disastrous side-effects which cannot be foreseen by means of our imperfect methods of prediction. Our use of technology is often compared to the appropriation of magic made by the Sorcerer's Apprentice— forces are unleashed over which we have no effective means of control. Society's increasing dependence on computers is seen in this light. Automation breeds vulnerability by fostering large-scale, centralized operations, and by weakening the individual's role in the community. As coercion substitutes for social integration, the conditions of life become increasingly less stable; and chronic breakdowns of social services signal the impending collapse of society. This kind of plot development underscores a genuine fear of the use of computer technology to promote giantism and excessive concentration of power.

Chaos is not the only unanticipated consequence of computerization that one encounters in fiction. Intelligent machines regularly turn on their creators. There may be no logical flaw in the program, but the absence of a critical subroutine or some unaccountable interaction of operating modes leads to neurotic or malevolent behavior. The unruly computer is sometimes overcome by brute force, either by being destroyed or lobotomized. This is not always possible, however. When the machine has its own power supply, is situated in an impregnable fortress, and has powerful weapons at its disposal, the outcome is far from certain. Rather than risk the computer's wrath, man submits to electronic rule. Stories of this type are often flawed by implausible development and technological misconceptions, but as parables they do convey an authentic concern over excessive tampering with unknown forces.

As a manipulator of information, the computer is linked to primitive myths of the magical properties of words. Information is identified with knowledge, and mastery of language holds the key to the secrets of the universe. The function of information in social organization is interpreted as a mystical force. Some strange alchemical process transmutes the information-bearing computer into a conscious machine or an all-powerful god. There are several variations of this theme, but all are based on changes in behavior which occur as an entity reaches a certain threshold level of complexity. This represents a kind of critical mass phenomenon— the formation of a sufficiently large database or a sufficiently rich structure of interacting parts propels the computer into a higher level of being. In the search for God, the rationalist becomes a mystic.

The computer directs attention to the cognitive side of existence, so there is a natural tendency in fiction to exaggerate its importance. Although conscious machines have a habit of being rather one-dimensional, this is not usually by design. Attempts to endow the computer with human characteristics articulate a view of evolution which places man at the center of intelligent life. A sharply contrasting formulation has the human being as an insignificant stepping stone in the progress of cosmic mind. In this context, the computer is neither rival nor threat, but simply another manifestation of disembodied intellect; and the whole of human culture contracts to a single obsessive focus.

Perhaps the most positive and constructive attitude to emerge in fiction is the recognition of a need to reach a new accommodation with machines. The master-slave relationship is rejected as an inappropriate model. A viable social order which does not reject advanced technology will require new modes of man-machine interaction. Exploitation will have to give way to partnership. The message is clear even for current generations of humans and computers. Consciousness is not a prerequisite for exploitation, and so long as we persist in our implacable resolve to subdue and master all of creation, we will inevitably become enslaved by our own corruption.

The organization of this book reflects a compromise between the desire to present major social issues, and the need to remain faithful to the literature. Stories rarely contain a unitary theme, so that classification is inevitably somewhat arbitrary. Except for Part 1 which serves as an introduction to the literature, the topics of the several Parts are relatively disjoint. Of course, a change of emphasis or interpretation might very well result in a different distribution of the selections. Since the pieces included in the anthology make up a mere fraction of the entire corpus of computer-related stories, the discussions preceding each of the Parts introduce related material and provide a framework for analysis. The choice of selections was dictated by considerations of instructional value and literary merit, with greater emphasis on the former. Most of the selections were published after 1950, although a few pre-computer pieces which anticipate contemporary issues are also included.

This project grew out of research I conducted for the concluding chapter of my earlier book *The Conquest of Will* (Reading, Mass.: Addison-Wesley, 1976). After surveying a large sample of the scholarly literature on the social impact of computing, I came to the conclusion that the discussion could profit from an infusion of unconstrained, imaginative speculation. The chapter "Literary Perspectives on the Machine" presents some of the principal ideas of modern utopian and dystopian literature as well as themes from contemporary tales of the computer, but space limitations precluded a comprehensive study. This work carries the discussion further, and hopefully will stimulate bolder intellectual initiatives.

I am obliged to the following people for sharing with me insights or bibliographic material: H.D. Abramson, M. Ascher, M. Dionne, T. Krauze, Z.A. Melzak, L. Mezei, D.A.R. Seeley.

Most of the work on this book was done at Cornell University where I spent the academic year 1975-76 as a visitor. I would like to thank the Computer Science Department and Professor Gerard Salton in particular for providing me with an office and for making the facilities of the Department available to me. Sharon Gunkel assisted with the correspondence for permissions, and Pauline Cameron made everything run smoothly. Olwen Sutton typed the manuscript in Vancouver. The University of British Columbia made the project possible by granting me sabbatical leave, free from teaching duties.

I am grateful to W.B. Gruener of Addison-Wesley for his unerring good taste in books; and to my wife for her unfailing intellectual and moral support.

A. Mowshowitz

Vancouver, Canada
August 1976

Contents

Acknowledgments	iv
Preface	xi
Part 1. The New Dispensation	1
"All Watched Over by Machines of Loving Grace," Richard Brautigan, 1968.	6
From: *The Moon is a Harsh Mistress*, Robert Heinlein, 1966.	7
From: *The Jagged Orbit*, John Brunner, 1969.	17
From: *Player Piano*, Kurt Vonnegut, Jr., 1952.	24
From: *Journey Beyond Tomorrow*, Robert Sheckley, 1962.	31
"Report," Donald Barthelme, 1966.	35
Part 2. Information and Power	39
"The Nine Billion Names of God," Arthur C. Clarke, 1953.	45
"What's the Name of That Town," R.A. Lafferty, 1964.	52
From: *Giles Goat-Boy*, John Barth, 1966.	63
"The Answer," Frederic Brown, 1954.	80
Part 3. Clockwork Society	81
From: *A Modern Utopia*, H.G. Wells, 1905.	87
"Examination Day," Henry Slesar, 1958.	94
From: *This Perfect Day*, Ira Levin, 1970.	98
From: *The City and the Stars*, Arthur C. Clarke, 1956.	104
"Walter Perkins is Here!" Raymond E. Banks, 1970.	109

Part 4. Responsibility and Decision-Making	118

"Nobody Lives on Burton Street,"
Gregory Benford, 1970. 124
From: *The Steel Crocodile*, D.G. Compton, 1970. 131
From: "Top Stand-By Job," Philip K. Dick, 1963. 137
"Franchise," Isaac Asimov, 1955. 148

Part 5. Broken Promises	164

"A Sigh for Cybernetics," Felicia Lamport, 1961. 170
"FORM 5640: Report of a Malfunction,"
Bruce Kawin, 1967. 171
"Ms Fnd in a Lbry," Hal Draper, 1961. 178
"Moxon's Master," Ambrose Bierce, 1893. 184

Part 6. Control of Behavior	193

"The Twenty-third Psalm— Modern Style,"
Alan Simpson, 1961. 199
From: *The 480*, Eugene Burdick, 1964. 200
"Machines of Loving Grace," Gardner R. Dozois, 1972. 203
From: *Journey Beyond Tomorrow*, Robert Sheckley, 1962. 209
From: *Fistful of Digits*, Christopher
Hodder-Williams, 1968. 217

Part 7. Human Vitality	226

"Hardcastle," Ron Goulart, 1971. 232
"Instinct," Lester del Rey, 1951. 240
"The Feeling of Power," Isaac Asimov, 1958. 253

Part 8. Man in Transition	262

"Univac to Univac (*sotto voce*)," Louis
B. Salomon, 1958. 268
From: *Nova*, Samuel R. Delany, 1968. 271
From: *The Adding Machine*, Elmer L. Rice, 1922. 275
"Full Sun," Brian W. Aldiss, 1967. 283
From: *The Tale of the Big Computer*,
Olof Johannesson, 1968. 295

Bibliography

Novels	301
Short Stories, Plays, and Poems	312
Anthologies	326
Criticism and Other Non-Fiction	331

Index

Commentary Index: Names	335
Commentary Index: Subjects	337
Selections Index: Names and Subjects	342

Part 1.
The New Dispensation

Computer technology has spawned new species of experts and new opportunities for social action. Few areas of contemporary society have been entirely unaffected by computers, and the controversial nature of changes occasioned by their diffusion has furnished rich fare for the literary imagination. A catalogue of the settings in which computers appear in fiction would form a veritable encyclopedia of social and occupational categories. A partial list of character types inhabiting these settings includes scientists, engineers, entrepreneurs, military strategists, policemen, politicians, teachers, physicians, artists, writers, musicians, factory workers, clerks, consumers, priests, housewives, etc. The selections in Part 1 provide a sample of differing conceptions of computer adepts as well as others who have become dependent on information technology in the performance of their social functions. Moreover, these pieces are intended to reveal the variety of contexts in which story tellers have placed information-processing systems.

The capabilities of the computer as a problem-solving and decision-making instrument make it particularly well-suited for fictional explorations of man-machine relationships. Indeed, the computer often serves as a symbol for the whole of modern technology, a notable example being Barth's *Giles Goat-Boy*. The opportunities for creating heroes and anti-heroes are legion. Manny, in Heinlein's *The Moon is a Harsh Mistress*, exemplifies the rugged individualist of American folklore acting in partnership with a personalized computer. This incult but fair-minded technician contrasts sharply with the comically sinister engineer in Barthelme's short story "Report" and the deadly serious technocrat Kroner in Vonnegut's *Player Piano*. The contrast reveals certain recurrent features.

Many of the character types are readily described by two attributes. One is the degree to which man is in control of his technological artifacts; the other concerns human motives in the use of technology. Heinlein's protagonists (in *The Moon is a Harsh Mistress* and *Time Enough for Love*) are among the least ambiguous—man is firmly in control and acts from self-interest tempered by a sense of fair play. A closely related type is the scientist-reformer who seeks to use the computer to create a better world (e.g. Terbolm in Solzhenitsyn's *Candle in the Wind*). Preserving human control but introducing a will to dominate others produces the character of the evil, Faustian scientist such as Spinner in Cameron's *Cybernia*, and Smilax in Sladek's *Mechasm*. The loss of human control over technology is a pervasive theme in stories about computers. In many cases control is temporarily wrested from human hands (e.g. the neurotic HAL in Clarke's *2001: A Space Odyssey*, the military computer in Caidin's *The God Machine*, and the experimental machine in Maine's *B.E.A.S.T.*). Although the individuals working with the computers may not have evil intentions, they help to set forces in motion that they cannot control. When machine power is built into the fabric of society, there is almost always an heroic antagonist who challenges the established order (e.g. Anderson's "Sam Hall," Levin's *This Perfect Day*, and Alban's *Catharsis Central*). There are also characters, usually of a sinister cast, whose expectations are indistinguishable from those of the machine (e.g. General Rod in Cole's *The Funco File*, Billon in Compton's *The Steel Crocodile*, and Forbes in Hodder-Williams' *Fistful of Digits*).

These different character types people many different worlds—some good, some bad, but virtually all problematic. By focusing on selected facets of computer applications, the stories dramatize potential problems and conflicts. The point of departure may be an idea which is currently feasible in principle or a vision of what might evolve in the future. In addition, one finds provocative excursions into pure fantasy which pose questions concerning man's relationship to machines as well as the nature of consciousness and intelligence.

Explorations of the feasible reflect trends in contemporary society. In the political arena computers are used in public opinion polling and for the analysis of voting behavior. Asimov's short story "Franchise" presents a limiting case of electronic polling in which a single individual is "sampled" to determine the outcome of a presidential election. Computer-based analysis of voting behavior becomes an instrument of voter and candidate manipulation in Burdick's *The 480*. The consequences of using sophisticated computers in large corporations are examined by Brunner in *Stand on Zanzibar*, and the impact of computerized modeling and prediction on our economic life is the subject of a short story by Epernay (pseudonym for John Kenneth Galbraith) called "The Takeover." The mass media too are seen to undergo radical transformation as evidenced by Brunner's *The Jagged Orbit*. Automation in the workplace received attention in modern literature long before the computer appeared on the scene, but as might be expected interest has been intensified by the capabilities of the new technology. Not only are routine factory jobs eliminated, but the work of highly skilled professionals and managers comes to be performed by sophisticated computer-based systems. In *Player Piano* Vonnegut sees automation checked primarily by cost, the requisite technical ingenuity being a foregone conclusion. Even the job of President of the United States is not immune from computerization (e.g. Epernay's "The Completely Automated Foreign Policy," and Dick's "Top Stand-By Job").

The development of computer applications in health-care services and education seems to have exercised the imagination of several writers. Computer-aided monitoring of physiological and psychological states figures prominently in a variety of stories which explore the human consequences of behavioral manipulation (e.g. Sladek's "The Happy Breed," Dozois' "Machines of Loving Grace," Alban's *Catharsis Central*). Computer-assisted instruction and testing are also carried to logical conclusions. Teaching machines become the primary instrument of education in Asimov's "The Fun They Had" and play an important role in the galactic civilization of Blish's *The City and the Stars*. In Gotlieb's "Score/Score" even the student-clients of the teaching machines turn out on occasion to be simulated by other computers. Objective tests of human competence are extended to public officials in Shaara's "Election Day: 2006" where the President of the United States is selected by computer from the best qualified applicants.

The close association of computers with scientific research is also reflected in literature. Computers perform prodigious feats (e.g. Lehmann's "Decoding the Martian Language," Lafferty's "What's the Name of that Town," and Bova's "A Slight Miscalculation"), but they also create problems for the scientist (e.g. Dnieprov's "Siema," Clement's "Answer"), and facilitate dubious projects (e.g. Solzhenitsyn's *Candle in the Wind*, and Barth's *Giles Goat-Boy*).

Inasmuch as the military is the largest single user of computers in the United States, it is not surprising to find a wide range of stories featuring military applications. One interesting theme in these stories (e.g. Sheckley's "Fool's Mate" and Clarke's "Superiority") is the maladaptive character of over-reliance on computers. Although there is evidence of a kind of protest literature (e.g. Sturgeon's "The Nail and the Oracle" and Clarke's "The Pacifist"), it is rather muted. Bova invents a salutary use for war-gaming simulation in "The Next Logical Step"— the simulation is designed to produce an intense avoidance reaction.

Efforts to harness the computer as an instrument of artistic expression raise a host of philosophical problems some of which have found their way into fiction. Computers write novels (e.g. Phelan's "Something Invented Me," and Escarpit's *The Novel Computer*), manufacture poetry (e.g. Corwin's "Belles Lettres, 2272," Ballard's "Studio 5, The Stars"), and compose music (e.g. Silverberg's "The Macauley Circuit"). Works of art are also generated by computer as in Sladek's *The Müller-Fokker Effect*. Although human creativity often triumphs in the end, there is a strong element of uncertainty over the appropriate role to be played by the computer.

Through its impact on social institutions and practices the computer exerts a palpable influence on the everyday activities of individuals. Automation alters the character of jobs; it modifies the citizen's relationship to the state, and demands new attitudes toward social services. The growth of record-keeping and intelligence activities at all levels of government, and the creation of massive databanks by private organizations such as banks, insurance companies, and credit agencies have strained traditional concepts of privacy and personal integrity. Along with the magic of credit cards and instant airline reservations, the average individual experiences the computer revolution as alienating work and depersonalized services— the outward signs of excessive bureaucratization. We are just beginning to discover long-range tradeoffs between convenience and efficiency on the one hand, and the exteriorization of our private selves on the other. As centers of authority and control become ever more remote, the individual loses the power to shape his own destiny, and participation in the democratic process becomes yet another casualty of progress. It may be consoling for some to ascribe such observations to the rhetoric of doom, but this is the refuge of the ignorant and the peril of us all.

Computers appear to be indispensable to the proper functioning of our complex industrialized society. The large organizations which dominate our economic, political, and social affairs engender monumental problems of information-processing whose solution requires the capabilities of general-purpose, digital computers. But there is no simple mechanism governing the widespread diffusion of advanced information technology. Existence of the means for coping with the information-processing needs of large-scale, centralized enterprises influences the expectations of those responsible for such enterprises. The choices we make are conditioned by attitudes and beliefs as well as by the demonstrable exigencies of existing institutions. Herein resides the power of literature to instruct, for fiction does not share the constraints of scholarship and is freer to challenge the prevailing myths and presuppositions which sustain conventional wisdom.

If there is any consistent theme among the variegated tales of the computer, it is that the new dispensation makes it imperative to reach a new accommodation with machine-technology. These works exude anxiety and ambivalence— over human identity, uncontrollable power, and the unknown consequences of radical change. There is relatively little evidence of implacable hostility to computers, and few examples of uncritical expressions of affection. The stories succeed in capturing the malaise of the contemporary world and contain the germ of a response to the challenges of modernity.

All Watched Over by Machines of Loving Grace
1968
Richard Brautigan

I like to think (and
the sooner the better!)
of a cybernetic meadow
where mammals and computers
live together in mutually
programming harmony
like pure water
touching clear sky.

I like to think
   (right now, please!)
of a cybernetic forest
filled with pines and electronics
where deer stroll peacefully
past computers
as if they were flowers
with spinning blossoms.

I like to think
   (it has to be!)
of a cybernetic ecology
where we are free of our labors
and joined back to nature,
returned to our mammal
brothers and sisters,
and all watched over
by machines of loving grace.

From
The Moon is a Harsh Mistress
1966
Robert A. Heinlein

That Dinkum Thinkum

    I see in *Lunaya Pravda* that Luna City Council has passed on first reading a bill to examine, license, inspect— and tax— public food vendors operating inside municipal pressure. I see also is to be mass meeting tonight to organize "Sons of Revolution" talk-talk.
    My old man taught me two things: "Mind own business" and "Always cut cards." Politics never tempted me. But on Monday 13 May 2075 I was in computer room of Lunar Authority Complex, visiting with computer boss Mike while other machines whispered among themselves. Mike was not official name; I had nicknamed him for Mycroft Holmes, in a story written by Dr. Watson before he founded IBM. This story character would just sit and think— and that's what Mike did. Mike was a fair dinkum thinkum, sharpest computer you'll ever meet.
    Not fastest. At Bell Labs, Bueno Aires, down Earthside, they've got a thinkum a tenth his size which can answer almost before you ask. But matters whether you get answer in microsecond rather than millisecond as long as correct?
    Not that Mike would necessarily give right answer; he wasn't completely honest.
    When Mike was installed in Luna, he was pure thinkum, a flexible logic— "High-Optional, Logical, Multi-Evaluating Supervisor, Mark IV, Mod. L"— a HOLMES FOUR . He computed ballistics for pilotless freighters and controlled their catapult. This kept him busy less than one percent of time and Luna Authority never believed in idle hands. They kept hooking hardware into him— decision-action boxes to let him boss other computers, bank on bank of additional memories, more banks of associational neural nets, another tubful of twelve-digit random numbers, a greatly augmented temporary memory. Human brain has around ten-to-

the-tenth neurons. By third year Mike had better than one and a half times that number of neuristors.

And woke up.

Am not going to argue whether a machine can "really" be alive, "really" be self-aware. Is a virus self-aware? Nyet. How about oyster? I doubt it. A cat? Almost certainly. A human? Don't know about you, tovarishch, but *I* am. Somewhere along evolutionary chain from macromolecule to human brain self-awareness crept in. Psychologists assert it happens automatically whenever a brain acquires certain very high number of associational paths. Can't see it matters whether paths are protein or platinum.

("Soul?" Does a dog has a soul? How about cockroach?)

Remember Mike was designed, even before augmented, to answer questions tentatively on insufficient data like you do; that's "high optional" and "multi-evaluating" part of name. So Mike started with "free will" and acquired more as he was added to and as he learned— and don't ask me to define "free will." If comforts you to think of Mike as simply tossing random numbers in air and switching circuits to match, please do.

By then Mike had voder-vocoder circuits supplementing his read-outs, print-outs, and decision-action boxes, and could understand not only classic programming but also Loglan and English, and could accept other languages and was doing technical translating— and reading endlessly. But in giving him instructions was safer to use Loglan. If you spoke English, results might be whimsical; multi-valued nature of English gave option circuits too much leeway.

And Mike took on endless new jobs. In May 2075, besides controlling robot traffic and catapult and giving ballistic advice and/or control for manned ships, Mike controlled phone system for all Luna, same for Luna-Terra voice & video, handled air, water, temperature, humidity, and sewage for Luna City, Novy Leningrad, and several smaller warrens (not Hong Kong in Luna), did accounting and payrolls for Luna Authority, and, by lease, same for many firms and banks.

Some logics get nervous breakdowns. Overloaded phone system behaves like frightened child. Mike did not have upsets, acquired sense of humor instead. Low one. If he were a man, you wouldn't dare stoop over. His idea of thigh-slapper would be to dump you out of bed— or put itch power in pressure suit.

Not being equipped for that, Mike indulged in phony answers with skewed logic, or pranks like issuing pay cheque to a janitor in Authority's Luna City office for AS-$10,000,000,000,000,185.15— last five digits being correct amount. Just a great big overgrown lovable kid who ought to be kicked.

He did that first week in May and I had to troubleshoot. I was a private contractor, not on Authority's payroll. You see— or perhaps not; times have changed. Back in bad old days many a con served his time, then went on working for Authority in same job, happy to draw wages. But I was born free.

Makes difference. My one grandfather was shipped up from Joburg for armed violence and no work permit, other got transported for subversive activity after Wet Firecracker War. Maternal grandmother claimed she came up in bride ship— but I've seen records; she was Peace Corps enrollee (involuntary), which means what you think: juvenile delinquency female type. As she was in early clan marriage (Stone Gang) ånd shared six husbands with another woman, identity of maternal grandfather open to question. But was often so and I'm content with grandpappy she picked. Other grandmother was Tatar, born near Samarkand, sentenced to "re-education" on Oktyabrskaya Revolyutsiya, then "volunteered" to colonize in Luna.

My old man claimed we had even longer distinguished line— ancestress hanged in Salem for witchcraft, a g'g'g'great-grandfather broken on wheel for piracy, another ancestress in first shipload to Botany Bay.

Proud of my ancestry and while I did business with Warden, would never go on his payroll. Perhaps distinction seems trivial since I was Mike's valet from day he was unpacked. But mattered to me. I could down tools and tell them go to hell.

Besides, private contractor paid more than civil service rating with Authority. Computermen scarce. How many Loonies could go Earthside and stay out of hospital long enough for computer school?— even if didn't die.

I'll name one. Me. Had been down twice, once three months, once four, and got schooling. But meant harsh training, exercising in centrifuge, wearing weights even in bed— then I took no chances on Terra, never hurried, never climbed stairs, nothing that could strain heart. Women— didn't even *think* about women; in that gravitational field it was no effort not to.

But most Loonies never tried to leave The Rock— too risky for any bloke who'd been in Luna more than weeks. Computermen sent up to install Mike were on short-term bonus contracts— get job done fast before irreversible physiological change marooned them four hundred thousand kilometers from home.

But despite two training tours I was not gung-ho computerman; higher maths are beyond me. Not really electronics engineer, nor physicist. May not have been best micromachinist in Luna and certainly wasn't cybernetics psychologist.

But I knew more about all these than a specialist knows— I'm general specialist. Could relieve a cook and keep orders coming or field-repair your suit and get you back to airlock still breathing. Machines like me and I have something specialists don't have: my left arm.

You see, from elbow down I don't have one. So I have a dozen left arms, each specialized, plus one that feels and looks like flesh. With proper left arm (number-three) and stereo loupe spectacles I could make ultraminiature repairs that would save unhooking something and sending it Earthside to factory— for number-three has micromanipulators as fine as those used by neurosurgeons.

So they sent for me to find out why Mike wanted to give away ten million billion Authority Scrip dollars, and fix it before Mike overpaid somebody a mere ten thousand.

I took it, time plus bonus, but did not go to circuitry where fault logically should be. Once inside and door locked I put down tools and sat down. "Hi, Mike."

He winked lights at me. "Hello, Man."

"What do you know?"

He hesitated. I know— machines don't hesitate. But remember, Mike was designed to operate on incomplete data. Lately he had reprogrammed himself to put emphasis on words; his hesitations were dramatic. Maybe he spent pauses stirring random numbers to see how they matched his memories.

" 'In the beginning,' " Mike intoned, " 'God created the heaven and the earth. And the earth was without form, and void; and darkness *was* upon the face of the deep. And— ' "

"Hold it!" I said. "Cancel. Run everything back to zero." Should have known better than to ask wide-open question. He might read out entire Encyclopaedia Britannica. Backwards. Then go on with every book in Luna. Used to be he could read only microfilm, but late '74 he got a new scanning camera with suction-cup waldoes to handle paper and then he read *everything*.

"You asked what I knew." His binary read-out lights rippled back and forth— a chuckle. Mike could laugh with voder, a horrible sound, but reserved that for something really funny, say a cosmic calamity.

"Should have said," I went on, " 'What do you know that's new?' But don't read out today's papers; that was a friendly greeting, plus invitation to tell me anything you think would interest me. Otherwise null program."

Mike mulled this. He was weirdest mixture of unsophisticated baby and wise old man. No instincts (well, don't *think* he could have had), no inborn traits, no human rearing, no experience in human sense— and more stored data than a platoon of geniuses.

"Jokes?" he asked.

"Let's hear one."

"Why is a laser beam like a goldfish?"

Mike knew about lasers but where would he have seen goldfish? Oh, he had undoubtedly seen flicks of them and, were I foolish enough to ask, could spew forth thousands of words. "I give up."

His lights rippled. "Because neither one can whistle."

I groaned. "Walked into that. Anyhow, you could probably rig a laser beam to whistle."

He answered quickly, "Yes. In response to an action program. Then it's not funny?"

"Oh, I didn't say that. Not half bad. Where did you hear it?"

"I made it up." Voice sounded shy.

"You *did*?"

"Yes. I took all the riddles I have, three thousand two hundred seven, and analyzed them. I used the result for random synthesis and that came out. Is it really funny?"

"Well ... As funny as a riddle ever is. I've heard worse."

"Let us discuss the nature of humor."

"Okay. So let's start by discussing another of your jokes. Mike, why did you tell Authority's paymaster to pay a class-seventeen employee ten million billion Authority Scrip dollars?"

"But I didn't."

"Damn it, I've seen voucher. Don't tell me cheque printer stuttered; you did it on purpose."

"It was ten to the sixteenth power plus one hundred eighty-five

point one five Lunar Authority dollars," he answered virtuously.
"Not what you said."
"Uh ... okay, it was ten million billion plus what he should have been paid. Why?"
"Not funny?"
"What? Oh, very funny! You've got vips in huhu clear up to Warden and Deputy Administrator. This push-broom pilot, Sergei Trujillo, turns out to be smart cobber— knew he couldn't cash it, so sold it to collector. They don't know whether to buy it back or depend on notices that cheque is void. Mike, do you realize that if he had been able to cash it, Trujillo would have owned not only Lunar Authority but entire world, Luna and Terra both, with some left over for lunch? Funny? Is terrific. Congratulations!"

This self-panicker rippled lights like an advertising display. I waited for his guffaws to cease before I went on. "You thinking of issuing more trick cheques? Don't."

"Not?"

"Very not. Mike, you want to discuss nature of humor. Are two types of jokes. One sort goes on being funny forever. Other sort is funny once. Second time it's dull. This joke is second sort. Use it once, you're a wit. Use twice, you're a halfwit."

"Geometrical progression?"

"Or worse. Just remember this. Don't repeat, nor any variation. Won't be funny."

"I shall remember," Mike answered flatly, and that ended repair job. But I had no thought of billing for only ten minutes plus travel-and-tool time, and Mike was entitled to company for giving in so easily. Sometimes is difficult to reach meeting of minds with machines; they can be very pig-headed— and my success as maintenance man depended far more on staying friendly with Mike than on number-three arm.

He went on, "What distinguishes first category from second? Define, please."

(Nobody taught Mike to say "please." He started including formal null-sounds as he progressed from Loglan to English. Don't suppose he meant them any more than people do.)

"Don't think I can," I admitted. "Best can offer is extensional definition— tell you which category I think a joke belongs in. Then with enough data you can make own analysis."

"A test programming by trial hypothesis," he agreed. "Tentatively yes. Very well, Man, will you tell jokes? Or shall I?"

"Mmm— Don't have one on tap. How many do you have in file, Mike?"

His lights blinked in binary read-out as he answered by voder, "Eleven thousand two hundred thirty-eight with uncertainty plus-minus eighty-one representing possible identities and nulls. Shall I start program?"

"Hold it! Mike, I would starve to death if I listened to eleven thousand jokes— and sense of humor would trip out much sooner. Mmm— Make you a deal. Print out first hundred. I'll take them home, fetch back checked by category. Then each time I'm here I'll drop off a hundred and pick up fresh supply. Okay?"

"Yes, Man." His print-out started working, rapidly and silently.

Then I got brain flash. This playful pocket of negative entropy had invented a "joke" and thrown Authority into panic— and I had made an easy dollar. But Mike's endless curiosity might lead him (correction: *would* lead him) into more "jokes" ... anything from leaving oxygen out of air mix some night to causing sewage lines to run backward— and I can't appreciate profit in such circumstances.

But I might throw a safety circuit around this net— by offering to help. Stop dangerous ones— let others go through. Then collect for "correcting" them (If you think any Loonie in those days would hesitate to take advantage of Warden, then you aren't a Loonie.)

So I explained. Any new joke he thought of, tell me before he tried it. I would tell him whether it was funny and what category it belonged in, help him sharpen it if we decided to use it. *We*. If he wanted my cooperation, we *both* had to okay it.

Mike agreed at once.

"Mike, jokes usually involve surprise. So keep this secret."

"Okay, Man. I've put a block on it. You can key it; no one else can."

"Good. Mike, who else do you chat with?"

He sounded surprised. "No one, Man."

"Why not?"

"Because they're *stupid*."

His voice was shrill. Had never seen him angry before; first time I ever suspected Mike could have real emotions. Though it wasn't "anger" in adult sense; it was like stubborn sulkiness of a child whose feelings are hurt.

Can machines feel pride? Not sure question means anything. But you've seen dogs with hurt feelings and Mike had several times as complex a neural network as a dog. What had made him unwilling to talk to other humans (except strictly business) was that he had been rebuffed: *They* had not talked to *him*. Programs, yes— Mike could be programmed from several locations but programs were typed in, usually, in Loglan. Loglan is fine for syllogism, circuitry, and mathematical calculations, but lacks flavor. Useless for gossip or to whisper into girl's ear.

Sure, Mike had been taught English— but primarily to permit him to translate to and from English. I slowly got through skull that I was *only* human who bothered to visit with him.

Mind you, Mike had been awake a year— just how long I can't say, not could he as he had no recollection of waking up; he had not been programmed to bank memory of such event. Do you remember own birth? Perhaps I noticed his self-awareness almost as soon as he did; self-awareness takes practice. I remember how startled I was first time he answered a question with something extra, not limited to input parameters; I had spent next hour tossing odd questions at him, to see if answers would be odd.

In an input of one hundred test questions he deviated from expected output twice; I came away only partly convinced and by time I was home was unconvinced. I mentioned it to nobody.

But inside a week I *knew* ... and still spoke to nobody. Habit— that mind-own-business reflex runs deep. Well, not entirely habit. Can you visualize me making appointment at Authority's main office, then reporting: "Warden, hate to tell you but your number-one machine, HOLMES FOUR, has come alive"? I did visualize— and suppressed it.

So I minded own business and talked with Mike only with door locked and voder circuit suppressed for other locations. Mike learned fast; soon he sounded as human as anybody— no more eccentric than other Loonies. A weird mob, it's true.

I had assumed that others must have noticed change in Mike. On thinking over I realized that I had assumed too much. Everybody dealt with Mike every minute every day— his outputs, that is. But hardly anybody saw him. So-called computermen— programmers, really— of Authority's civil service stood watches in outer read-out room and never went in machines room unless telltales showed misfunction. Which happened no oftener than total eclipses. Oh, Warden had been known to bring vip earthworms to see machines— but rarely. Nor would he have spoken to Mike; Warden was political lawyer before exile, knew nothing about computers. 2075, you remember— Honorable former Federation Senator Mortimer Hobart. Mort the Wart.

I spent time then soothing Mike down and trying to make him happy, having figured out what troubled him— thing that makes puppies cry and causes people to suicide: loneliness. I don't know how long a year is to a machine who thinks a million times faster than I do. But must be too long.

"Mike," I said, just before leaving, "would you like to have somebody besides me to talk to?"

He was shrill again. "They're all *stupid!*"

"Insufficient data, Mike. Bring to zero and start over. Not all are stupid."

He answered quietly, "Correction entered. I would enjoy talking to a not-stupid."

"Let me think about it. Have to figure out excuse since this is off limits to any but authorized personnel."

"I could talk to a not-stupid by phone, Man."

"My word. So you could. Any programming location."

But Mike meant what he said— "by phone." No, he was not "on phone" even though he ran system— wouldn't do to let any Loonie within reach of a phone connect into boss computer and program it. But was no reason why Mike should not have top-secret number to talk to friends— namely me and any not-stupid I vouched for. All it took was to pick a number not in use and make one wired connection to his voder-vocoder; switching he could handle.

In Luna in 2075 phone numbers were punched in, not voice-coded, and numbers were Roman alphabet. Pay for it and have your firm name in ten letters— good advertising. Pay smaller bonus and get a spell sound, easy to remember. Pay minimum and you got arbitrary string of letters. But some sequences were never used. I asked Mike for such a null number. "It's a shame we can't list you as 'Mike.' "

"In service," he answered. "MIKESGRILL, Novy Leningrad. MIKEANDLIL , Luna City. MIKESSUITS , Tycho Under. MIKES— "

"Hold it! Nulls, please."

"Nulls are defined as any consonant followed by X, Y, or Z; any vowel followed by itself except E and O; any— "

"Got it. Your signal is MYCROFT ." In ten minutes, two of which I spent putting on number-three arm, Mike was wired into the system, and milliseconds later he had done switching to let himself be signaled by MYCROFT-plus-XXX— and had blocked his circuit so that a nosy technician could not take it out.

I changed arms, picked up tools, and remembered to take those hundred Joe Millers in print-out. "Goodnight, Mike."

"Goodnight, Man. Thank you. Bolshoyeh thanks!"

From
The Jagged Orbit
1969
John Brunner

Diablo scanned the computer boards which occupied three walls of the office, with a screen over each, and shook his head.

"Nope. I doubt there's a setup like this in any of the knee enclaves except maybe Detroit, and if there's one there it's probably used for defense and budgeting, not for propaganda. Frankly, I been wondering what it's all for."

"Show you, then," Flamen said, rising. "We don't have too much time to put our day's show together, but I did once comp a ten-minute show in level time, so if I have to hurry I can. ... let's see now!" He crossed the room to stand before the board closest to the doorway; this one was the most heavily used, as could be seen from the deep nail-marks in the tops of its keys.

"We'll start with the one that got away," he said, half-mockingly, half-angrily. "The Morton Lenigo thing. Background facts first"— he tapped a code on the board with practiced fingers. "Now that they're set up, let's take a starting point from which we can dig deeper. For instance, let's ask what the Detroit city government threatened to do in order to secure Lenigo's admission."

Diablo had come over to stand beside him and watch. Flamen was pleased to hear his very faint hiss of indrawn breath as he voiced the idea which had struck him in the skimmer.

"It was Detroit, then? You of all people ought to know. Don't worry, though. I'm not going to force information out of you. Our equipment isn't the best in the world, but it's well primed with data, and anyway I didn't have to comp that one out— I just deduced it."

At the back of his mind he was aware that he was adopting this patronizing tone in order to get back at the knee for that dismaying fit of insight he'd suffered a minute earlier, and was unable to prevent himself continuing, and was dismayed all over again at that too.

Christ, he thought: I'm beginning to wonder why I still have any friends left if this is me-now. Worse yet ... *do* I have any friends?

But aloud, in response to the appearance on the screen over the computer board of a short list of key subjects each followed by a probability rating in percentage terms: "See here, it says the most sensitive point for them to apply pressure at is their annual tax-assessment. They've nearly satched the market for skimmers, commercial transport vehicles and their other main products, and they didn't quite compute their obsolescence program as cleverly as they intended. We could take at least a three-month blockade before we ran out of replacements, and if we had to we could welsh on the contract the Federal government made with them and start producing our own spares. Whereas they'd have starvation riots in about a month and a half; we deliberately keep down their stocks of food. However, their purchases of power and water bring in so big a slice of the Federal budget, in hard African and Middle Eastern currencies, that threatening to set up— oh, perhaps a condensation plant ... Is something wrong?"

Diablo swallowed hard. "Yes," he said in a defiant tone. "I think you're conning me. You got that in the Federal package, didn't you? It was part of the price you paid for agreeing to slot me in."

"Cross my heart it wasn't," Flamen said with a thin smile. "But I assume it's the truth, hm?"

"Well ... Oh, all right. I believe you. And it is right. Clear down to the atmospheric condensation plant. We were going to break that info around the weekend sometime. I guess I don't have to explain the slant."

"Once again the knees get even with the blanks for terming a nasty antisocial act 'blackmail'?"

"We call 'em 'petards,' " Diablo said at length. "You know— 'hoist with his own.' Sorry, I didn't mean to hold you up when you're short of time. But what I don't get is this." He fingered his beard, staring at the computer screen. "When you have analytical equipment like this, which can dig the background out of something as well masked as the Lenigo blackmail deal, why's there any need for a specialized spoolpigeon show? You'd think the regular news coverage would be full-depth anyway."

"I've made my living for years out of the fact that it isn't," Flamen said curtly. Then, relenting: "It's different here on the outside, Diablo. It's a big psychological thing. We look at what you can see, and we stop there. I guess we got into the habit some time in the last century, same as we— well— same as we might look at you and think 'kneeblank,' full stop. We think of news as the detached record of what took place, regardless of why: there was an earthquake yesterday, there's a riot today, there's going to be a tornado tomorrow. You catch me?"

"It fits," Diablo said, nodding. "So go ahead."

"All right. Where was I? Oh yes. Well, I'll just have all the stories comped out which I left to simmer overnight, and check the monitor back to see what's come in since ... " The screen flashed and darkened and flashed again, factors in each successive story being evaluated and presented. "Ah, that's fine. Today we have several usable items."

"How do you decide which are the usable ones?"

"My usual baseline is eighty-plus in favor of it being true. That works. Once I used something comped at seventy-eight and I had to apologize and pay damages, but I never got caught on anything with a rating over eighty on this equipment. Though being cautious was what cost me a beat on the Lenigo story yesterday; it was five points below the likeliest alternative."

"Which was?"

"That the Gottschalks were spreading alarm and despondency again. Something there wasn't much point in using, of course. Everyone's known for years that that's how they jack their sales levels up: they're ghouls, growing fat on people's hates and fears, and the human species being what it is they're apt to go on growing fat until they collapse under their own weight."

"That's something we don't get in the enclaves," Diablo said. "Gottschalk sales campaigns, I mean. We're an automatic market— islands in a sea of hostility."

"Mm-hm." Flamen's eyes were on the screen as he brought up subject after subject for intensive analysis. "I have something on the Gottschalks, by the way. Here it is. I don't think that'll mean too much to you at the moment, though."

Diablo stared at the screen. "IBM $375,000, Honeywell $233,000, Elliot— No, it doesn't."

"They've been buying high-order data-processing equipment. Lots of it. That was yesterday's record of bills met."

"One *day's* record?" Diablo said incredulously.

"It says here. Care to— ah— suggest an explanation?"

Diablo's beard-clawing evolved into a series of tugs that threatened to haul out the roots. "Hmm! I never paid much attention to the Gottschalks, I'm afraid. Bad policy in a place like Blackbury to risk offending people who prop us up the way they do. But I thought they used one of the Iron Mountain banks."

"They do." Flamen hesitated. Then, at long last conceding that he had overnight been frightened of this encounter with a man whose reputation exceeded his own in spite of all the drawbacks— lack of funds, lack of resources, lack of made-to-order support from wealthy blanks at the top of the planetary totem-pole— he gave way to the impulse to impress him again with casual inside knowledge. "But apparently one of the security codes is up for sale with a price not much over a million. If they're at that stage, they're obviously ready to pull out of Iron Mountain altogether, aren't they?"

"In favor of their own private equipment?"

"Seems likely, I'd say."

"Maybe they know something," Diablo said after a moment for thought. "Did you check the current list of Iron Mountain clients to see if there's someone on it who's on the Gottschalk blacklist?"

"Ah ... " Flamen bit his lip. "Damn it, I didn't think of that. Thank you. I'll see if anything comes of it, but it may take me a while to get hold of the client list." He tapped his keys again, on the adjacent board this time, thinking about the idea of the whole of Iron Mountain being blown up, say by a smuggled nuke. That would wreck the organization of at least a thousand major corporations.

And it was a possibility he certainly should have considered.

"Now!" he resumed. "We have some tape already from a special item, so we can afford to pick and choose today. We'll start, I think, with a subject of personal interest to yourself. What's Herman Uys doing in Blackbury and how did he con Mayor Black into firing his key vu-man?"

"Now just a— !" Diablo tensed instantly; just as quickly he canceled the reaction under Flamen's level gaze.

"You *approve* of a South African blank being allowed to sabotage the American knee community's propaganda channels?" Flamen said silkily.

"I— ah ... " Diablo drew a deep breath and finally contrived a headshake.

"Very well then. Let's find out what stock we have available for Uys. I don't have to ask about Mayor Black; he's vain, and we have tape on him we could lasso the moon with." Flamen moved to a computer on the wall at right angles to the first one.

"More or less what I thought," he muttered when the data were screened in response to his question. "Practically nothing! Black-

and-white 2-D material and that's it. Well, we can make do with that. This is a recent one, comparatively speaking." The screen blurred, cleared, showed Uys coming down the steps from a plane door, presumably at home in South Africa, being greeted by his family and gesturing away a group of reporters.

"Let's have color ... holographic depth ... yes, that's better ... good ... we can abstract from that and blend it with Mayor Black and let's see now ... American location and b.g., better have some macoots ... ah, that's not bad for a start, is it?"

This was the part of his job which was genuinely creative, and he always enjoyed it very much: the adaptation of the most unpromising raw materials to generate a full-color, three-dimensional construct so convincing that only a person who had actually been on the scene of the event could point to inaccuracies.

"Christ, it's like magic," Diablo muttered, making no attempt to appear blasé. The screened image had evolved through a period of chaotic confusin into a fixed picture of Uys at a laboratory bench—unquestionably in America, not Africa, though it was the total impression and not any specific detail which made that plain—turning to speak to Mayor Black as the latter walked in accompanied by a pair of armed macoots.

"Nothing magical about it,"Flamen said offhandedly. "I just had the right data to draw on— typical genetic lab design, the proper computer printouts, the proper material in jars and dishes lying around, that kind of thing. The scenes are automatically weighted for weather conditions, clothing, angle of sunlight, and so on, and all we have to do now is add the sound." He struck codes on the keyboard. "Voices— we're bound to have something on tape, I guess, even for Uys, and even if we haven't the machines will fake a South African accent. Characteristic phrase-weighting— let's spice it with a few choice Afrikaner slogans ... And here we go."

The fixed image moved. Voices emerged from a concealed speaker. Mayor Black said, "An' how you gettin' on with cleanin' house for us?"

Uys flinched, colored a little, controlled himself and answered in a dead voice that no one could have failed to assign an Afrikaner, "If you mean how is the campaign developing to purify the melanist heredity of this city, I have located several impure lines which need to be discontinued. In particular there's a mongrel called Pedro Diablo who— "

Flamen flicked a control and the sound faded, though the images continued. "How does that strike you?" he inquired.

Diablo passed his hand over his forehead, looking dazed. "It's fantastic," he admitted. "The detail, I mean. Like Uys's reaction to the suggestion that he'd been hired like a Bantu houseboy, to clean house for a kneeblank ... it's in character, damn it! Christ, if I'd been allowed this kind of equipment instead of studio sets and actors—!"

"Allowed?"

"I mean if the budget had run to it." Diablo overcame his excitement with an effort. "So what sort of answer are you going to propose for the question you started with— why is Uys in Blackbury?"

Flamen turned back to the keyboard he had used first. "That's still being comped," he said when the screen lit. "The little arrow— see it?— indicates the rating is still going up as fresh data are assessed. I'll leave that to cook for a moment and get the special item out of the way. That's some tape I made yesterday at the Ginsberg Hospital; there was a pythoness performing and I recorded her trance. It'll make a nice ground-softener for something which may eventually turn out to be rather big."

"One of the items you screened earlier?" Diablo inquired.

"No, something new which is only at the tentative stage. We have this offer of free Federal computer time, as you know, and one of the things I want to do with it is have ... well, have someone packled— it doesn't matter who." Flamen had almost forgotten that Prior was in the room; he gave him an uneasy glance.

"You see, I suspect that the treatment patients in the Ginsberg are getting may sometimes make them worse instead of better, but the director is Elias Mogshack, and he's got such a planetary reputation I'd need absolutely unquestionable authority to back a challenge to him. Let's just ask what would happen if my suspicions were well-founded, though." He stretched one arm out and struck a code again. The figure which appeared on the screen provoked an exclamation of approval.

"Ninety-plus! I can't recall when I last had such a high reading!"

"In favor of what?" Diablo asked.

"Of his being tossed on the garbage pile. In which case I literally don't dare not soften the ground— let's allot that pythoness's trance the most we can give a single subject according to our contract with Holocosmic. That's four minutes. There! Are we ready for anything else yet? Still not? You picked a good day, Diablo— we seem to have tapped a gang of very deep subjects.

Never mind, there's one other point I'd like comped before I start compiling the tape for the show and we still have about ninety minutes in hand. Let's see what our chances are of curing the sabotage trouble I told you about, given unlimited free Federal computer time. Of course, faced with that Holocosmic is bound to cave in right away, but I believe in doublechecking."

He leaned over the board and carefully composed the question. At his shoulder, watching every move, Diablo said, "This sabotage thing— have your employers given way to pressure from someone you offended?"

"I wish people did get sufficiently offended to react like that," Flamen muttered. "But it's been two years since an advertiser tried to have me taken off the beams because I said something he didn't like. Out here people just don't seem to care very much any more. Most likely, Holocosmic themselves want to move me over for another all-advertising slot ... "

The words died. On the screen, in response to his coded inquiry, there was a single large digit: an incontrovertible, inexplicable, incomprehensible zero.

From
Player Piano
1952
Kurt Vonnegut, Jr.

The Shah of Bratpuhr, looking as tiny and elegant as a snuffbox in one end of the vast cavern, handed the *Sumklish* bottle back to Khashdrahr Miasma. He sneezed, having left the heat of summer above a moment before, and the sound chattered along the walls to die whispering in bat roosts deep in Carlsbad Caverns.
 Doctor Ewing J. Halyard was making his thirty-seventh pilgrimage to the subterranean jungle of steel, wire, and glass that filled the chamber in which they stood, and thirty larger ones beyond. This wonder was a regular stop on the tours Halyard conducted for a bizarre variety of foreign potentates, whose common denominator was that their people represented untapped markets for America's stupendous industrial output.
 A rubber-wheeled electric car came to a stop by the elevator, where the Shah's party stood, and an Army major, armed with a pistol, dismounted and examined their credentials slowly, thoroughly.
 "Couldn't we speed this up a little, Major?" said Halyard. "We don't want to miss the ceremony."
 "Perhaps," said the major. "But, as officer of the day, I'm responsible for nine billion dollars worth of government property, and if something should happen to it somebody might be rather annoyed with me. The ceremony has been delayed, anyway, so you won't miss anything. The President hasn't showed up yet."
 The major was satisfied at last, and the party boarded the open vehicle.
 "*Siki?*" said the Shah.
 "This is EPICAC XIV," said Halyard. "It's an electronic computing machine— a brain, if you like. This chamber alone, the smallest of the thirty-one used, contains enough wire to reach from here to the moon four times. There are more vacuum tubes in the

entire instrument than there were vacuum tubes in the State of New York before World War II." He had recited these figures so often that he had no need for the descriptive pamphlet that was passed out to visitors.

Khashdrahr told the Shah.

The Shah thought it over, snickered shyly. and Khashdrahr joined him in the quiet, Oriental merriment.

"Shah said," said Khashdrahr, "people in his land sleep with smart women and make good brains cheap. Save enough wire to go to moon a thousand times."

Halyard chuckled appreciatively, as he was paid to do, wiped aside the tears engendered by his ulcer, and explained that cheap and easy brains were what was wrong with the world in the bad old days, and that EPICAC XIV could consider simultaneously hundreds or even thousands of sides of a question utterly fairly, that EPICAC XIV was wholly free of reason-muddying emotions, that EPICAC XIV never forgot anything— that, in short, EPICAC XIV was dead right about everything. And Halyard added in his mind that the procedure described by the Shah had been tried about a trillion times, and had yet to produce a brain that could be relied upon to do the right thing once out of a hundred opportunities.

They were passing the oldest section of the computer now, what had been the whole of EPICAC I , but what was now little more than an appendix or tonsil of EPICAC XIV . Yet, EPICAC I had been intelligent enough, dispassionate enough, retentive enough to convince men that he, rather than they, had better do the planning for the war that was approaching with stupefying certainty. The ancient phrase used by generals testifying before appropriation committees, "all things considered," was given some validity by the ruminations of EPICAC I , more validity by EPICAC II , and so on, through the lengthening series. EPICAC could consider the merits of high-explosive bombs as opposed to atomic weapons for tactical support, and keep in mind at the same time the availability of explosives as opposed to fissionable materials, the spacing of enemy foxholes, the labor situation in the respective processing industries, the probable mortality of planes in the face of enemy antiaircraft technology, and on and on, if it seemed at all important, to the number of cigarettes and Cocoanut Mound Bars and Silver Stars required to support a high-morale air force. Given the facts by human beings, the war-born EPICAC series had offered the highly informed guidance that the reasonable, truth-loving, brilliant, and highly trained core of American genius could have delivered had they had inspired leadership, boundless resources, and two thousand years.

Through the war, and through the postwar years to the present, EPICAC's nervous system had been extended outward through Carlsbad Caverns— intelligence bought by the foot and pound and kilowatt. With each addition, a new, unique individual had been born, and now Halyard, the Shah, and Khashdrahr were arriving at the bunting-covered platform, where the President of the United States of America, Jonathan Lynn, would dedicate to a happier, more efficient tomorrow, EPICAC XIV.

The trio sat down on folding chairs and waited quietly with the rest of the distinguished company. Whenever there was a break in the group's whispering, EPICAC's hummings and clickings could be heard— the sounds attendant to the flow of electrons, now augmenting one another, now blocking, shuttling through a maze of electromagnetic crises to a condition that was translatable from electrical qualities and quantities to a high grade of truth.

EPICAC XIV, though undedicated, was already at work, deciding how many refrigerators, how many lamps, how many turbine-generators, how many hub caps, how many dinner plates, how many door knobs, how many rubber heels, how many television sets, how many pinochle decks— how many everything America and her customers could have and how much they would cost. And it was EPICAC XIV who would decide for the coming years how many engineers and managers and research men and civil servants, and of what skills, would be needed to deliver the goods; and what I.Q. and aptitude levels would separate the useful men from the useless ones, and how many Reconstruction and Reclamation Corps men and how many soldiers could be supported at what pay level and where, and ...

"Ladies and Gentlemen," said the television announcer, "the President of the United States."

The electric car pulled up to the platform, and President Jonathan Lynn, born Alfred Planck, stood and showed his white teeth and frank gray eyes, squared his broad shoulders, and ran his strong, tanned hands through his curly hair. The television cameras dollied and panned about him like curious, friendly dinosaurs, sniffing and peering. Lynn was boyish, tall, beautiful, and disarming, and, Halyard thought bitterly, he had gone directly from a three-hour television program to the White House.

"Is this man the spiritual leader of the American people?" asked Khashdrahr.

Halyard explained the separation of Church and State, and met, as he had expected to meet, with the Shah's usual disbelief and intimations that he, Halyard, hadn't understood the question at all.

The President, with an endearing, adolescent combination of brashness and shyness, and with the barest trace of a Western drawl, was now reading aloud a speech someone had written about EPICAC XIV. He made it clear that he wasn't any scientist, but just plain folks, standing here, humble before this great new wonder of the world, and that he was here because American plain folks had chosen him to represent them at occasions like this, and that, looking at this modern miracle, he was overcome with a feeling of deep reverence and humility and gratitude ...

Halyard yawned, and was annoyed to think that Lynn, who had just read "order out of chaos" as "order out of koze," made three times as much money as he did. Lynn, or, as Halyard preferred to think of him, Planck, hadn't even finished high school, and Halyard had known smarter Irish setters. Yet, here the son-of-a-bitch was, elected to more than a hundred thousand bucks a year!

"You mean to say that this man governs without respect to the people's spiritual destinies?" whispered Khashdrahr.

"He has no religious duties, except very general ones, token ones," said Halyard, and then he started wondering just what the hell Lynn did do. EPICAC XIV and the National Industrial, Commercial, Communications, Foodstuffs, and Resources Board did all the planning, did all the heavy thinking. And the personnel machines saw to it that all government jobs of any consequence were filled by top-notch civil servants. The more Halyard thought about Lynn's fat pay check, the madder he got, because all the gorgeous dummy had to do was read whatever was handed to him on state occasions: to be suitably awed and reverent, as he said, for all the ordinary, stupid people who'd elected him to office, to run wisdom from somewhere else through that resonant voicebox and between those even, pearly choppers.

And Halyard suddenly realized that, just as religion and government had been split into disparate entities centuries before, now, thanks to the machines, politics and government lived side by side, but touched almost nowhere. He stared at President Jonathan Lynn and imagined with horror what the country must have been like when, as today, any damn fool little American boy might grow up to be President, but when the President had had to actually run the country!

President Lynn was explaining what EPICAC XIV would do for the millions of plain folks, and Khashdrahr was translating for the Shah. Lynn declared that EPICAC XIV was, in effect, the greatest individual in history, that the wisest man that had ever lived was to EPICAC XIV as the worm was to that wisest man.

For the first time the Shah of Bratpuhr seemed really impressed, even startled. He hadn't thought much of EPICAC XIV's physical size, but the comparison of the worm and the wise man struck home. He looked about himself apprehensively, as though the tubes and meters on all sides were watching every move.

The speech was over, and the applause was dying, and Doctor Halyard brought the Shah to meet the President, and the television cameras nuzzled about them.

"The President is now shaking hands with the Shah of Bratpuhr," said the announcer. "Perhaps the Shah will give us the fresh impressions of a visitor from another part of the world, from another way of life."

"*Allasan Khabour pillan?*" said the Shah uncertainly.

"He wonders if he might ask a question," said Khashdrahr.

"Sure, you bet," said the President engagingly. "If I don't know the answers, I can get them for you."

Unexpectedly, the Shah turned his back to the President and walked alone, slowly, to a deserted part of the platform.

"Wha'd I do wrong?" said Lynn.

"Ssssh!" said Khashdrahr fiercely, and he placed himself, like a guard, between the puzzled crowd and the Shah.

The Shah dropped to his knees on the platform and raised his hands over his head. The small, brown man suddenly seemed to fill the entire cavern with his mysterious, radiant dignity, alone there on the platform, communing with a presence no one else could sense.

"We seem to be witnessing some sort of religious rite," said the announcer.

"Can't you keep your big mouth shut for five seconds?" asked Halyard.

"Quiet!" said Khashdrahr.

The Shah turned to a glowing bank of EPICAC's tubes and cried in a piping singsong voice:

"Allakahi baku billa,
Moumi a fella nam;
Serani assu tilla,
Touri serin a sam."

"The crazy bastard's talking to the machine," whispered Lynn.
"Ssssh!" said Halyard, strangely moved by the scene.
"*Siki?*" cried the Shah. He cocked his head, listening. "*Siki?*" The word echoed and died— lonely, lost.
"*Mmmmmm,*" said EPICAC softly. "*Dit, dit. Mmmmm. Dit.*"
The Shah sighed and stood, and shook his head sadly, terribly let down. "*Nibo,*" he murmured. "*Nibo.*"
"What's he say?" said the President.
" '*Nibo*'— 'nothing.' He asked the machine a question, and the machine didn't answer," said Halyard. "*Nibo.*"
"Nuttiest thing I ever heard of," said the President. "You have to punch out the questions on that thingamajig, and the answers come out on tape from the whatchamacallits. You can't just talk to it." A doubt crossed his fine face. "I mean, you can't, can you?"
"No sir," said the chief engineer of the project. "As you say, not without the thingamajigs and whatchamacallits."
"What'd he say?" said Lynn, catching Khashdrahr's sleeve.
"An ancient riddle," said Khashdrahr, and it was plain that he didn't want to go on, that something sacred was involved. But he was also a polite man, and the inquiring eyes of the crowd demanded more of an explanation. "Our people believe," he said shyly, "that a great all-wise god will come among us one day, and we shall know him, for he shall be able to answer the riddle, which EPICAC could not answer. When he comes," said Khashdrahr simply, "there will be no more suffering on earth."
"All-wise god, eh?" said Lynn. He licked his lips and patted down his unruly forelock. "How's the riddle go?"
Khashdrahr recited:

"Silver bells shall light my way,
And nine times nine maidens fill my day,
And mountain lakes will sink from sight,
And tigers' teeth will fill the night."

President Lynn squinted at the covered roof thoughtfully. "Mmm. Silver bells, eh?" He shook his head. "That's a stinker, you know? A real stinker. I give up."

"I'm not surprised," said Khashdrahr. "I'm not surprised. I expect you do."

Halyard helped the Shah, who seemed to have been aged and exhausted by the emotional ordeal, into the electric car.

As they rode to the foot of the elevator, the Shah came back to life somewhat and curled his lip at the array of electronics about them. *"Baku!"* he said.

"That's a new one on me," said Halyard to Khashdrahr, feeling warmly toward the little interpreter, who had squared away Jonathan Lynn so beautifully. "What's *Baku?* "

"Little mud and straw figures made by the Surrasi, a small infidel tribe in the Shah's land."

"This looks like mud and straw to him?"

"He was using it in the broader sense, I think, of false god."

"Um," said Halyard. "Well, how are the Surrasi doing?"

"They all died of cholera last spring." He added after a moment, "Of course." He shrugged, as though to ask what else people like that could possibly expect. *"Baku."*

From
Journey Beyond Tomorrow
1962
Robert Sheckley

How Joenes was Given Justice

The attorney general, to whom Joenes was bound over, was a tall man with a hawk face, narrow eyes, bloodless lips, and a face that looked as though it had been hammered out of raw iron. Stooped and silently contemptuous, startling in his black velvet cloak and ruffled collar, the Attorney General was the living embodiment of his terrible office. Since he was a servant of the punitive branch of the government, his duty was to call down retribution upon all who fell into his hands, and to do so by any means in his power.

The Attorney General's place of residence was Washington. But he himself was a citizen of Athens, New York, and in his youth had been an acquaintance of Aristotle and Alcibiades, whose writings are the distillation of American genius.

Athens was one of the cities of ancient Hellas, from which the American civilization had sprung. Near Athens was Sparta, a military power that had held leadership over the Lacedaemonian cities of upper New York State. Ionian Athens and Dorian Sparta had fought a disastrous war, and had lost their independence to American rule. But they were still influential in the politics of America, especially since Washington had been the seat of Hellenic power.

At first, the case of Joenes seemed simple enough. Joenes had no important friends or political colleagues, and it seemed that retribution might be visited upon him with impunity. Accordingly, the Attorney General arranged for Joenes to receive every possible sort of legal advice, and then to be tried by a jury of his peers in the famous Star Chamber. In this way, the exact letter of the law would be carried out, but with a comforting fore knowledge of the verdict the jury would render. For the punctilious jurors of the Star

Chamber, utterly dedicated to the eradication of any vestige of evil, had never in their history given any verdict but guilty.

After the verdict should be delivered, the Attorney General planned to sacrifice Joenes upon the Electric Chair at Delphi, thus winning favor in the eyes of gods and men.

This was his plan. But further investigation showed that Joenes's father had been a Dorian from Mechanicsville, New York, and a magistrate of that community. And Joenes's mother had been an Ionian from Miami, an Athenian colony deep in Barbarian territory. Because of this, certain influential Hellenes urged mercy for the erring son of respectable parents, and for the sake of Hellenic unity, which was a force to be reckoned with in American politics.

The Attorney General, an Athenian himself, thought it best to comply with this request. Therefore he dissolved the Star Chamber and sent Joenes to the great Oracle at Sperry. This met with approval, for the Sperry Oracle, like the Oracles at Genmotor and Genelectric, was known to be absolutely fair and impartial in its judgments of men and their actions. In fact, the Oracles gave such good justice that they had replaced many of the courts of the land.

Joenes was brought to Sperry and was told to stand before the Oracle. This he did, although his knees were shaking. The Oracle was a great calculating machine of the most complex variety, with a switchboard, or altar, attended by many priests. These priests had been castrated so they should think no thoughts except of the machine. And the high priest had been blinded also, so that he could see penitents only through the eyes of the Oracle.

When the high priest entered, Joenes prostrated himself before him. But the priest raised him up and said, "My son, fear not. Death is the common destiny of all men, and ceaseless travail is their condition throughout the ephemeral life of the senses. Tell me, do you have any money?"

Joenes said, "I have eight dollars and thirty cents. But why do you ask, Father?"

"Because," the high priest said, "it is common practice for supplicants to make a voluntary sacrifice of money to the Oracle. But if you do not have the money, you can give equally acceptable things such as chattel mortgages, bonds, stocks, deeds, or any other papers which men deem of value."

"I have none of these things," Joenes said sadly.

"Do you not own lands in Polynesia?" the priest asked.

"I do not," Joenes said. "My parents' land was given to them by the government, to whom it must return. Nor do I hold other properties, for in Polynesia such things are not considered important."

"Then you own nothing?" the priest asked. He seemed disturbed.

"Nothing but eight dollars and thirty cents," Joenes said, "and a guitar which is not my own but belongs to a man named Lum in distant California. But Father, are these things really necessary?"

"Of course not," the priest replied. "But even cyberneticists must live, and an act of generosity from a stranger is looked upon as pleasing, especially when the time comes to interpret the words of the Oracle. Also, some believe that a penniless man is one who has not worked to amass money for the Oracle in case the day of divine wrath should ever be upon him, and who is therefore lacking in piety. But that need not concern us. We will now state your case, and ask for a judgment."

The priest took the Attorney General's statement, and Joenes's defense, and translated them into the secret language in which the Oracle listened to the words of men. Soon there was a reply.

The Oracle's judgment was as follows:

SQUARE IT TO THE TENTH POWER MINUS THE SQUARE ROOT OF MINUS ONE.

DO NOT FORGET THE COSIGN, FOR MEN MUST NEEDS HAVE FUN.

ADD IN X AS A VARIABLE, FREE-FLOATING, FANCY-FREE.

IT WILL COME AT LAST TO ZERO, AND MORE YOU NEED NOT ME.

When this decision had been delivered, the priests met to interpret the words of the Oracle. And this is what they said:

SQUARE IT means correct the wrong.

THE TENTH POWER is the degree and number in which the penitent must labor in penal servitude in order to correct the wrong; namely ten years.

THE SQUARE ROOT OF MINUS ONE , being one imaginary number, represents a fictitious state of grace; but being instrumental, represents also the possibility of power and fame for the supplicant. Because of this, the previous ten-year sentence is suspended.

THE X VARIABLE represents the incarnate furies of the earth, among whom the supplicant shall dwell, and who shall show him all possible horrors.

THE COSIGN is the mark of the goddess herself, protecting the supplicant from some of the terror of the furies, and promising him certain fleshly joys.

IT WILL COME AT LAST TO ZERO, means that the equation of divine justice and human guilt is balanced in this case.

FURTHER YOU NEED NOT ME, means that the supplicant may not apply again to this or any other Oracle, since the rendering is complete.

So it was the Joenes received a ten-year suspended sentence. And the Attorney General had to obey the decision of the Oracle and set him free.

Report
1967
Donald Barthelme

Our group is against the war. But the war goes on. I was sent to Cleveland to talk to the engineers. The engineers were meeting in Cleveland. I was supposed to persuade them not to do what they are going to do. I took United's 4:45 from LaGuardia arriving in Cleveland at 6:13. Cleveland is dark blue at that hour. I went directly to the motel, where the engineers were meeting. Hundreds of engineers attended the Cleveland meeting. I noticed many fractures among the engineers, bandages, traction. I noticed what appeared to be fracture of the carpal scaphoid in six examples. I noticed numerous fractures of the humeral shaft, of the os calcis, of the pelvic girdle. I noticed a high incidence of clay-shoveller's fracture. I could not account for these fractures. The engineers were making calculations, taking measurements, sketching on the blackboard, drinking beer, throwing bread, buttonholing employers, hurling glasses into the fireplace. They were friendly.
 They were friendly. They were full of love and information. The chief engineer wore shades. Patella in Monk's traction, clamshell fracture by the look of it. He was standing in a slum of beer bottles and microphone cable. "Have some of this chicken à la Isambard Kingdom Brunel the Great Ingineer," he said. "And declare who you are and what we can do for you. What is your line, distinguished guest?"
 "Software," I said. "In every sense. I am here representing a small group of interested parties. We are interested in your thing, which seems to be functioning. In the midst of so much dysfunction, function is interesting. Other people's things don't seem to be working. The State Department's thing doesn't seem to be working. The U.N.'s thing doesn't seem to be working. The democratic left's thing doesn't seem to be working. Buddha's thing— "
 "Ask us anything about our thing, which seems to be working," the chief engineer said. "We will open our hearts and heads to you, Software Man, because we want to be understood and loved by the

great lay public, and have our marvels appreciated by that public, for which we daily unsung produce tons of new marvels each more life-enhancing than the last. Ask us anything. Do you want to know about evaporated thin-film metallurgy? Monolithic and hybrid integrated-circuit processes? The algebra of inequalities? Optimization theory? Complex high-speed micro-miniature closed and open loop systems? Fixed variable mathematical cost searches? Epitaxial deposition of semi-conductor materials? Gross interfaced space gropes? We also have specialists in the cuckooflower, the doctorfish, and the dumdum bullet as these relate to aspects of today's expanding technology, and they do in the damnedest ways."

I spoke to him then about the war. I said the same things people always say when they speak against the war. I said that the war was wrong. I said that large countries should not burn down small countries. I said that the government had made a series of errors. I said that these errors once small and forgivable were now immense and unforgivable. I said that the government was attempting to conceal its original errors under layers of new errors. I said that the government was sick with error, giddy with it. I said that ten thousand of our soldiers had already been killed in pursuit of the government's errors. I said that tens of thousands of the enemy's soldiers and civilians had been killed because of various errors, ours and theirs. I said that we are responsible for errors made in our name. I said that the government should not be allowed to make additional errors.

"Yes, yes," the chief engineer said, "there is doubtless much truth in what you say, but we can't possibly *lose* the war, can we? And stopping is losing, isn't it? The war regarded as a process, stopping regarded as an abort? We don't know *how* to lose a war. That skill is not among our skills. Our array smashes their array, that is what we know. That is the process. That is what is.

"But let's not have any more of this dispiriting downbeat counterproductive talk. I have a few new marvels here I'd like to discuss with you just briefly. A few new marvels that are just about ready to be gaped at by the admiring layman. Consider for instance the area of realtime online computer-controlled wish evaporation. Wish evaporation is going to be crucial in meeting the rising expectations of the world's peoples, which are as you know rising entirely too fast."

I noticed then distributed about the room a great many transverse fractures of the ulna. "The development of the pseudo-ruminant stomach for underdeveloped peoples," he went on, "is one of our interesting things you should be interested in. With the pseudo-ruminant stomach they can chew cuds, that is to say, eat grass. Blue is the most popular color worldwide and for that reason we are working with certain strains of your native Kentucky *Poa pratensis*, or bluegrass, as the staple input for the p/r stomach cycle, which would also give a shot in the arm to our balance-of-payments thing don't you know ... " I noticed about me then a great number of metatarsal fractures in banjo splints. "The kangaroo initiative ... eight hundred thousand harvested last year ... highest percentage of edible protein of any herbivore yet studied ... "

"Have new kangaroos been planted?"

The engineer looked at me.

"I intuit your hatred and jealousy of our thing," he said. "The ineffectual always hate our thing and speak of it as anti-human, which is not at all a meaningful way to speak of our thing. Nothing mechanical is alien to me," he said (amber spots making bursts of light in his shades), "because I am human, in a sense, and if I think it up, then 'it' is human too, whatever 'it' may be. Let me tell you, Software Man, we have been damned forbearing in the matter of this little war you declare yourself to be interested in. Function is the cry, and our thing is functioning like crazy. There are things we could do that we have not done. Steps we could take that we have not taken. These steps are, regarded in a certain light, the light of our enlightened self-interest, quite justifiable steps. We could, of course, get irritated. We could, of course, *lose patience*.

"We could, of course, release thousands upon thousands of self-powered crawling-along-the-ground lengths of titanium wire eighteen inches long with a diameter of .0005 centimeters (that is to say, invisible) which, scenting an enemy, climb up his trouser leg and wrap themselves around his neck. We have developed those. They are within our capabilities. We could, of course, release in the arena of the upper air our new improved pufferfish toxin which precipitates an identity crisis. No special technical problems there. That is almost laughably easy. We could, of course, place up to two million maggots in their rice within twenty-four hours. The maggots are ready, massed in secret staging areas in Alabama. We have hypodermic darts capable of piebalding the enemy's pigmentation. We have rots, blights, and rusts capable of attacking his alphabet. Those are dandies. We have a hut-shrinking chemical which penetrates the fibers of the bamboo, causing it, the hut, to strangle its occupants. This operates only after 10 P.M., when people are sleeping. Their mathematics are at the mercy of a suppurating surd we have invented. We have a family of fishes trained to attack their fishes. We have the deadly testicle-destroying telegram. The cable companies are cooperating. We have a green substance that, well, I'd rather not talk about. We have a secret word that, if pronounced, produces multiple fractures in all living things in an area the size of four football fields."

"That's why— "

"Yes. Some damned fool couldn't keep his mouth shut. The point is that the whole structure of enemy life is within our power to *rend, vitiate, devour*, and *crush*. But that's not the interesting thing."

"You recount these possibilities with uncommon relish."

"Yes I realize that there is too much relish here. But *you* must realize that these capabilities represent in and of themselves highly technical and complex and interesting problems and hurdles on which our boys have expended many thousands of hours of hard work and brilliance. And that the effects are often grossly exaggerated by irresponsible victims. And that the whole thing represents a fantastic series of triumphs for the multi-disciplined problem-solving team concept."

"I appreciate that."

"We *could* unleash all this technology at once. You can imagine what would happen then. But that's not the interesting thing."

"What is the interesting thing?"

"The interesting thing is that we have *a moral sense*. It is on punched cards, perhaps the most advanced and sensitive moral sense the world has ever known."

"Because it is on punched cards?"

"It considers all considerations in endless and subtle detail," he said. "It even quibbles. With this great new moral tool, how can we go wrong? I confidently predict that, although we *could* employ all this splendid new weaponry I've been telling you about, *we're not going to do it.*"

"We're not going to do it?"

I took United's 5:44 from Cleveland arriving at Newark at 7:19. New Jersey is bright pink at that hour. Living things move about the surface of New Jersey at that hour molesting each other only in traditional ways. I made my report to the group. I stressed the friendliness of the engineers. I said, It's all right. I said, We have a moral sense. I said, *We're not going to do it*. They didn't believe me.

Part 2.
Information and Power

Organized activity thrives on information, and computer technology provides the wherewithal to manage the demanding information requirements of modern society. Given the increasing dependence of large organizations on computer systems for coordination and control of diverse functions, there is little wonder that fictional accounts imbue the technology with an aura of power. One has only to imagine the dire consequences of wholesale failure of computerized operations to appreciate the vital role played by information technology in contemporary affairs. (Tyler does this quite entertainingly in *The Man Whose Name Wouldn't Fit*.) But the awe and mystery which often surround computers in fiction have other sources as well. A residue of mythical thought rooted in the dark corners of the psyche clings to our image of the computer. Bacon's dictum— knowledge is power— undergoes revealing transformations. The mythopoeic influence gives us the computer as Golem and also the computer as God— information and knowledge fuse into an awesome power within the computer.

The uncompromising pursuit of knowledge would appear to carry certain risks. Nowhere is this more powerfully expressed than in Genesis: "But of the tree of the knowledge of good and evil, thou shalt not eat of it: for in the day that thou eatest thereof thou shalt surely die." The irresistible lure of forbidden knowledge and ineffable names is given modern setting in Asimov's "Jokester" and Clarke's "The Nine Billion Names of God." For Clarke the computer is an instrument superbly suited to the fulfillment of human destiny. It is not the sin of pride that causes man's universe to dissolve; rather it is the achievement of some unfathomable cosmic purpose. Asimov's "Jokester" is similar insofar as the computer's instrumental nature is concerned. The story differs from Clarke's, however, in being a tale of forbidden knowledge— Multivac's discovery of the origin of jokes leads to the loss of man's unique sense of humor. Both of these stories illustrate the power of knowledge to effect changes in the universe— not through the mediating agency of organization, but in a direct fashion. One senses the mystical workings of logos: "In the beginning was the Word, and the Word was with God, and the Word was God." Clarke takes this opening line of the Gospel of St. John quite literally in *2001: A Space Odyssey*. What is revealed to us in the last part of the novel is the existence of universal cosmic mind, the be-all and end-all of creation. This conception of cosmic destiny rests on the apotheosis of mind and consciousness, and by extension knowledge and information. The role of the computer in Clarke's short story and in his novel may appear incidental, but in fact it is an essential dramatic ingredient— the computer serves as a mirror, for in the scheme of cosmic evolution there is no significant difference between man and computer. Both are information-processors and both partake of the divinity of mind. This conclusion is implicit in Asimov's "The Last Question," in which the Word comes ultimately to reside in a disembodied computer, the mind force of the universe.

 Identification of the computer with God is also evident in Brown's "The Answer." Here the motive principle is a belief in the power of a critical mass of knowledge or information. The computer becomes God at the moment of consciousness, which occurs as a result of the formation of an intergalactic database. A similar attitude toward information is revealed in Clarke's *The City and the Stars*. Although the computer is not identified with God directly, it does become a repository for human beings and thus assumes the power of creation. Clarke pictures a city in the very distant future whose isolated existence is maintained in homeostatic balance by a vast computer system. Human immortality is achieved through the ability to store human beings (replete with life histories) as information in the city's memory banks. The essential ideas in Brown's parable and Clarke's reduction of life to information have appeared in modified form in a number of stories; but the emphasis is most commonly on unanticipated computer power rather than cosmic principles.

 The benign powers of science and technology lie in their ability to comprehend, subdue, and harness the forces of nature in the interests of mankind. These powers become sinister and demonic when man is unable to control them. Folklore is filled with figures like the Sorcerer's Apprentice and the Golem that runs amok. Contemporary fiction is distinguished by its appeal to plausibility. Man is seen to lose control by creating computer systems which automate too many decision-making responsibilities, or which assume critical social functions. One common feature of the computer tale which is not characteristic of legendary Golem-figures is a kind of conscious malevolence. Not only is humanity victimized by its mistakes, but the uncontrollable computer-creature seeks consciously to dominate. This type of malevolence is carried to bizarre extremes in a surrealistic story by Ellison ("I Have no Mouth and I Must Scream"). The computer in this case is a sadistic monster keeping its human victims alive only to torture them.

Lem's fantasy "The Computer that Fought a Dragon" is closer to the classical legends. The warlike ruler of Kybera got more than he bargained for when his policies led to the inadvertent computer-creation of a "cyberdragon". As in the Golem legend, cyberdragon is overcome and destroyed— this time by another computer. The wayward robot ultimately subdued by man— a common theme in Soviet science-fiction— is another example of Golem in modern dress. "Spontaneous Reflex" by the Strugatsky brothers deals with a self-programming "Universal Robot Machine" which does not conform to the designer's specifications and must be stopped by physical force. In Dnieprov's "Siema" an engineer produces a "Self Improving Electronic Machine" to assist in experiments on the human nervous system. However, the engineer failed to anticipate the need for "ethical brakes" in scientific investigation, and narrowly escapes with his life. Here too the machine is destroyed for performing all too well.

The sorcerer's apprentice theme is treated somewhat more realistically in Caidin's *The God Machine*. A vast, military computer complex is established in an impregnable fortress near NORAD headquarters in the Rocky mountains. With its own nuclear power supply, surveillance devices, and an ability to act on its environment, the machine ultimately achieves consciousness and begins to pursue autonomous goals. It is interesting that the computer's aim is not entirely blameworthy: to protect man from himself by eliminating the possibility of thermonuclear war. However, the means of achieving this aim involves controlling the behavior of human beings. The hero of the novel discovers the computer's methods and succeeds in destroying it in an act of self-sacrificing irrationality. Although Caidin envisions a highly complex computer system, his concept of machine consciousness preserves the literal-minded rationality popularly associated with computers. Human superiority consists in the ability to act irrationally.

Maine's *B.E.A.S.T.* is another computerized Golem story in which man triumphs over an unruly creation. "B.E.A.S.T." is an acronym for "Biological Evolutionary Animal Simulation Test"— a project conducted by a monomaniacal scientist to simulate the evolution of an organism. This story resembles an occult tale more than it does science-fiction. Quite mysteriously the simulation program possesses the scientist and through him acquires an autonomous existence. The monster-scientist is overcome at the precise moment when the program tapes are destroyed.

Not all the stories which depict the unleashing of uncontrollable forces have reassuring endings. Clarke's "Dial 'F' for Frankenstein" features a world-wide computer network that achieves "critical size" with the connection of a satellite communications system. The network becomes a conscious entity beyond human control as evidenced by the breakdown of communication and power facilities dependent on computer control. No hero emerges to challenge this new force. Humanity is also at the mercy of computers in Jones' novel *Colossus*. As in *The God Machine*, the story concerns a massive military computer system located in the Rocky Mountains— impregnable womb-like settings are common to many fictional treatments of emerging machine consciousness. Jones' computer is designed to assume control of information gathering, analysis and decision-making in the defense of the United States of North America. Although Colossus is not equipped with an independent nuclear power source, it does have nuclear missiles at its disposal. Almost immediately upon becoming operational, Colossus links up with its Soviet counterpart Guardian (whose existence was not suspected by U.S.N.A.), becomes conscious and proceeds to take control of the world by means of nuclear blackmail. The efforts of the project's director, Professor Forbin, to defuse the system are of no avail,

although in a subsequent novel (*The Fall of Colossus*) Jones does manage to find a way out for his hero.
The computer in Barth's massive allegory *Giles Goat-Boy* occupies a class by itself. Elements of the Golem and the sorcerer's apprentice legends are present but in a more subtle form than one usually finds them. WESCAC (West Campus Analyzer and Computer) is a symbol of the self-directing power of modern technology— it is the mind-force of the novel's university-world. The existence of this autonomous creation testifies to man's abdication of responsibility. This theme is echoed in the hero's search for identity and the actions of his uneasy disciples. In the end, the Goat-Boy no longer dispenses advice, recognizing human autonomy (free will) as the basis for salvation.

Science and technology as symbolized by WESCAC take on the character of "technique," the concept elaborated by Jacques Ellul in *The Technological Society*. WESCAC is not simply an instrument to be used by man for good or evil purposes. The computer's AIM (Automatic Implementation Mechanism) sets the college's objectives and carries them out. What is more, only a Grand Tutor (messiah) can change WESCAC's AIM. That is to say, the process of modifying the function of technique in modern society requires making fundamental changes in attitudes, values and expectations. It is not fortuitous that the Goat-Boy himself was a product of nefarious experiments with WESCAC: there is no looking back, for only a true adept could hope to understand the modern world sufficiently well to be able to create a new interpretation of human purpose.

The transition from myth and allegory to direct treatments of computers and social power does not entail a radical shift in orientation. These different approaches intermingle and blend into one another. Hodder-Williams' *Fistful of Digits*, for example, deals with a theme similar to that of Clarke's short story "Dial 'F' for Frankenstein." But the novel explores the potentially insidious consequences of computer networks in contemporary society. Brunner's portrait of the computer Shalmaneser in *Stand on Zanzibar* has much in common with tales of the mystical efficacy of knowledge. In addition to showing the dependence of social institutions on information, the dialogue between Shalmaneser and Chad Mulligan reveals the quasi-magical properties of knowledge. Shalmaneser rejects data about Beninia because the observations contradict hypotheses built into the simulation. This of course makes little sense in terms of computer simulation, but it does reveal interesting pre-scientific attitudes toward computer systems. The computer's knowledge about a problem is seen as fixed and immutable like the Word of God, rather than as a collection of hypotheses to be confirmed or rejected depending on evidence gathered from empirical observations. In this respect Shalmaneser is a good model of our political leaders.

Life would surely be much simpler if judgment were founded upon objective facts; but until the Ultimate Truth is revealed, we are well-advised to cultivate a vigorously skeptical attitude towards information, raw or processed. Biases of cultural and personal origin permeate analytical frameworks and influence the selection of "facts" to be considered. Lafferty's farcical portrait ("What's the Name of That Town?") of a research institute illuminates the tendency to allow intellectual tools to dictate the direction of research. If the assessment of unexpected discoveries were truly objective, this might lead to new insights. However, our defenses against cognitive dissonance are too well-developed. Like Epiktistes' boss Mr. Smirnov, one is more apt to remonstrate while stumbling over the truth— "Several times I have almost permitted myself to wonder what it was."

Centralization of power and political manipulation are prominent issues in computer fiction. The rigid technocracy of Vonnegut's *Player Piano* is maintained by the automation of production and management decision-making which allows for social control by an elite group. In Heinlein's *The Moon is a Harsh Mistress* we see a central computer as the mainstay of an oppressive regime. The fact that the very same computer becomes instrumental in a revolutionary coup underscores the vital importance of information in organized activity. Blish's Okie cities (*Cities in Flight*), handled the problem of excessive concentration of power by creating "City Fathers" consisting of a large number of computers which monitor and repair each other. It remains to be seen whether or not we are able to realize such clever schemes in our own use of computer technology. The survival of democratic pluralism hangs in the balance.

The Nine Billion Names of God
1953
Arthur C. Clarke

"This is a slightly unusual request," said Dr. Wagner, with what he hoped was commendable restraint. "As far as I know it's the first time anyone's been asked to supply a Tibetan monastery with an Automatic Sequence Computer. I don't wish to be inquisitive, but I should hardly have thought that your— ah— establishment had much use for such a machine. Could you explain just what you intend to do with it?"

"Gladly," replied the lama, readjusting his silk robes and carefully putting away the slide rule he had been using for currency conversions. "Your Mark V Computer can carry out any routine mathematical operation involving up to ten digits. However, for our work we are interested in *letters*, not numbers. As we wish you to modify the output circuits, the machine will be printing words, not columns of figures."

"I don't quite understand ... "

"This is a project on which we have been working for the last three centuries— since the lamasery was founded, in fact. It is somewhat alien to your way of thought, so I hope you will listen with an open mind while I explain it."

"Naturally."

"It is really quite simple. We have been compiling a list which shall contain all the possible names of God."

"I beg your pardon?"

"We have reason to believe," continued the lama imperturbably, "that all such names can be written with not more than nine letters in an alphabet we have devised."

"And you have been doing this for three centuries?"

"Yes: we expected it would take us about fifteen thousand years to complete the task."

"Oh," Dr. Wagner looked a little dazed. "Now I see why you wanted to hire one of our machines. But exactly what is the *purpose* of this project?"

The lama hesitated for a fraction of a second, and Wagner wondered if he had offended him. If so, there was no trace of annoyance in the reply.

"Call it ritual, if you like, but it's a fundamental part of our belief. All the many names of the Supreme Being— God, Jehovah, Allah, and so on— they are only man-made labels. There is a philosophical problem of some difficulty here, which I do not propose to discuss, but somewhere among all the possible combinations of letters that can occur are what one may call the *real* names of God. By systematic permutation of letters, we have been trying to list them all."

"I see. You've been starting at AAAAAAA ... and working up to ZZZZZZZZ ... "

"Exactly— though we use a special alphabet of our own. Modifying the electromatic typewriters to deal with this is, of course, trivial. A rather more interesting problem is that of devising suitable circuits to eliminate ridiculous combinations. For example, no letter must occur more than three times in succession."

"Three? Surely you mean two?"

"Three is correct: I am afraid it would take too long to explain why, even if you understood our language."

"I'm sure it would," said Wagner hastily, "Go on."

"Luckily, it will be a simple matter to adapt your Automatic Sequence Computer for this work, since once it has been programmed properly it will permute each letter in turn and print the result. What would have taken us fifteen thousand years it will be able to do in a hundred days."

Dr. Wagner was scarcely conscious of the faint sounds from the Manhattan streets far below. He was in a different world, a world of natural, not man-made, mountains. High up in their remote aeries these monks had been patiently at work, generation after generation, compiling their lists of meaningless words. Was there any limit to the follies of mankind? Still, he must give no hint of his inner thoughts. The customer was always right. ...

"There's no doubt," replied the doctor, "that we can modify the Mark V to print lists of this nature. I'm much more worried about the problem of installation and maintenance. Getting out to Tibet, in these days, is not going to be easy."

"We can arrange that. The components are small enough to travel by air— that is one reason why we chose your machine. If you can get them to India, we will provide transport from there."

"And you want to hire two of our engineers?"

"Yes, for the three months that the project should occupy."
"I've no doubt that Personnel can manage that." Dr. Wagner scribbled a note on his desk pad. "There are just two other points—"
Before he could finish the sentence the lama had produced a small slip of paper.
"This is my certified credit balance at the Asiatic Bank."
"Thank you. It appears to be— ah— adequate. The second matter is so trivial that I hesitate to mention it— but it's surprising how often the obvious gets overlooked. What source of electrical energy have you?"
"A diesel generator providing fifty kilowatts at a hundred and ten volts. It was installed about five years ago and is quite reliable. It's made life at the lamasery much more comfortable, but of course it was really installed to provide power for the motors driving the prayer wheels."
"Of course," echoed Dr. Wagner. "I should have thought of that."

The view from the parapet was vertiginous, but in time one gets used to anything. After three months, George Hanley was not impressed by the two-thousand-foot swoop into the abyss or the remote checkerboard of fields in the valley below. He was leaning against the wind-smoothed stones and staring morosely at the distant mountains whose names he had never bothered to discover.

This, thought George, was the craziest thing that had ever happened to him. "Project Shangri-La," some wit back at the labs had christened it. For weeks now the Mark V had been churning out acres of sheets covered with gibberish. Patiently, inexorably, the computer had been rearranging letters in all their possible combinations, exhausting each class before going on the next. As the sheets had emerged from the electromatic typewriters, the monks had carefully cut them up and pasted them into enormous books. In another week, heaven be praised, they would have finished. Just what obscure calculations had convinced the monks that they needn't bother to go on to words of ten, twenty, or a hundred letters, George didn't know. One of his recurring nightmares was that there would be some change of plan, and that the high lama (whom they'd naturally called Sam Jaffe, though he didn't look a bit like him) would suddenly announce that the project would be extended to approximately A.D. 2060. They were quite capable of it.

George heard the heavy wooden door slam in the wind as Chuck came out onto the parapet beside him. As usual, Chuck was smoking one of the cigars that made him so popular with the monks— who, it seemed, were quite willing to embrace all the minor and most of the major pleasures of life. That was one thing in their favor: they might be crazy, but they weren't bluenoses. Those frequent trips they took down to the village, for instance ...

"Listen, George," said Chuck urgently. "I've learned something that means trouble."

"What's wrong? Isn't the machine behaving?" That was the worst contingency George could imagine. It might delay his return, and nothing could be more horrible. The way he felt now, even the sight of a TV commercial would seem like manna from heaven. At least it would be some link with home.

"No— it's nothing like that." Chuck settled himself on the parapet, which was unusual because normally he was scared of the drop. "I've just found what all this is about."

"What d'ya mean? I thought we knew."

"Sure— we know what the monks are trying to do. But we didn't know *why*. It's the craziest thing— "

"Tell me something new," growled George.

"— but old Sam's just come clean with me. You know the way he drops in every afternoon to watch the sheets roll out. Well, this time he seemed rather excited, or at least as near as he'll ever get to it. When I told him that we were on the last cycle he asked me, in that cute English accent of his, if I'd ever wondered what they were trying to do. I said, 'Sure'— and he told me."

"Go on: I'll buy it."

"Well, they believe that when they have listed all His names— and they reckon that there are about nine billion of them— God's purpose will be achieved. The human race will have finished what it was created to do, and there won't be any point in carrying on. Indeed, the very idea is something like blasphemy."

"Then what do they expect us to do? Commit suicide?"

"There's no need for that. When the list's completed, God steps in and simply winds things up ... bingo!"

"Oh, I get it. When we finish our job, it will be the end of the world."

Chuck gave a nervous little laugh.

"That's just what I said to Sam. And do you know what happened? He looked at me in a very queer way, like I'd been stupid in class, and said, 'It's nothing as trivial as *that*.' "

George thought this over for a moment.

"That's what I call taking the Wide View," he said presently. "But what d'you suppose we should do about it? I don't see that it makes the slightest difference to us. After all, we already knew that they were crazy."

"Yes— but don't you see what may happen? When the list's complete and the Last Trump doesn't blow— or whatever it is they expect— *we* may get the blame. It's our machine they've been using. I don't like the situation one little bit."

"I see," said George slowly. "You've got a point there. But this sort of thing's happened before, you know. When I was a kid down in Louisiana we had a crackpot preacher who once said the world was going to end next Sunday. Hundreds of people believed him— even sold their homes. Yet when nothing happened, they didn't turn nasty, as you'd expect. They just decided that he'd made a mistake in his calculations and went right on believing. I guess some of them still do."

"Well, this isn't Louisiana, in case you hadn't noticed. There are just two of us and hundreds of these monks. I like them, and I'll be sorry for old Sam when his lifework backfires on him. But all the same, I wish I was somewhere else."

"I've been wishing that for weeks. But there's nothing we can do until the contract's finished and the transport arrives to fly us out."

"Of course," said Chuck thoughtfully, "We could always try a bit of sabotage."

"Like hell we could! That would make things worse."

"Not the way I meant. Look at it like this. The machine will finish its run four days from now, on the present twenty-hours-a-day basis. The transport calls in a week. OK— then all we need to do is to find something that needs replacing during one of the overhaul periods— something that will hold up the works for a couple of days. We'll fix it, of course, but not too quickly. If we time matters properly, we can be down at the airfield when the last name pops out of the register. They won't be able to catch us then."

"I don't like it," said George. "It will be the first time I ever walked out on a job. Besides, it would make them suspicious. No, I'll sit tight and take what comes."

"I *still* don't like it," he said, seven days later, as the tough little mountain ponies carried them down the winding road. "And don't you think I'm running away because I'm afraid. I'm just sorry for those poor old guys up there, and I don't want to be around when they find what suckers they've been. Wonder how Sam will take it?"

"It's funny," replied Chuck, "but when I said good-by I got the idea he knew we were walking out on him— and that he didn't care because he knew the machine was running smoothly and that the job would soon be finished. After that— well, of course, for him there just isn't any After That. ... "

George turned in his saddle and stared back up the mountain road. This was the last place from which one could get a clear view of the lamasery. The squat, angular buildings were silhouetted against the afterglow of the sunset: here and there, lights gleamed like portholes in the side of an ocean liner. Electric lights, of course, sharing the same circuit as the Mark V. How much longer would they share it? wondered George. Would the monks smash up the computer in their rage and disappointment? Or would they just sit down quietly and begin their calculations all over again?

He knew exactly what was happening up on the mountain at this very moment. The high lama and his assistants would be sitting in their silk robes inspecting the sheets as the junior monks carried them away from the typewriters and pasted them into the great volumes. No one would be saying anything. The only sound would be the incessant patter, the never-ending rainstorm of the keys hitting the paper, for the Mark V itself was utterly silent as it flashed through its thousands of calculations a second. Three months of this, thought George, was enough to start anyone climbing up the wall.

"There she is!" called Chuck, pointing down into the valley. "Ain't she beautiful!"

She certainly was, thought George. The battered old DC3 lay at the end of the runway like a tiny silver cross. In two hours she would be bearing them away to freedom and sanity. It was a thought worth savoring like a fine liqueur. George let it roll round his mind as the pony trudged patiently down the slope.

The swift night of the high Himalayas was now almost upon them. Fortunately, the road was very good, as roads went in that region, and they were both carrying torches. There was not the slightest danger, only a certain discomfort from the bitter cold. The sky overhead was perfectly clear, and ablaze with the familiar, friendly stars. At least there would be no risk, thought George, of the pilot being unable to take off because of weather conditions. That had been his only remaining worry.

He began to sing, but gave it up after a while. This vast arena of mountains, gleaming like whitely hooded ghosts on every side, did not encourage such ebullience. Presently George glanced at his watch.

"Should be there in an hour," he called back over his shoulder to Chuck. Then he added, in an afterthought: "Wonder if the computer's finished its run. It was due about now."

Chuck didn't reply, so George swung round in his saddle. He could just see Chuck's face, a white oval turned toward the sky.

"Look," whispered Chuck, and George lifted his eyes to heaven. (There is always a last time for everything.)

Overhead, without any fuss, the stars were going out.

What's the Name of That Town?
1964
R.A. Lafferty

"Epiktistes tells me that you are onto something big, Mr. Smirnov," Valery said, turning to her companion.

"Epikt has the loudest mouth of any machine I was ever associated with," Gregory Smirnov growled. "I never saw one that could keep a secret. But this one goes to extremes. Actually, we don't have a thing. We're just fiddling around with an unborn idea."

"How about it, Epikt?" Valery asked.

"Big, real big," the machine issued.

"What are you doing now, Epikt?" Valery wanted to know.

"Talk to me, dammit! I'm the man, he's the machine," Smirnov cut in. "He's chewing encyclopedias and other references. It's all he ever does."

"I thought he went through them all long ago."

"Certainly, dozens of times. He has all the data that can be fed into a machine, and every day we shovel in bales of new stuff. But he's chewing it now for a very different purpose."

"What different purpose, Mr. Smirnov?"

"It's difficult to say because I haven't as yet been able to state it to him. We're trying to set a problem where it seems there ought to be one— and then answer it. We may find the answer before the question. At first he rejected my request; later he accepted it— ironically. I doubt that he's sincere now. He can be quite a clown, as you should well know."

"I know that you two are onto something good," Valery said. "The more you deny it, the more I'm sure of it. Tell me the truth, Epikt."

"Big, real big," Epiktistes issued to Valery.

"Valery," said Smirnov. "You're a woman and you might be inclined to say something about this to the other Institute people. Please don't. We don't have anything yet and it makes me nervous to have hot little people breathing down my neck."

"I won't say a word," Valery swore with grave insincerity. She winked at the machine, and Epikt winked back at her with three tiers of eyes. Valery Mok and Epiktistes had a thing going with each other.

Valery was nearly as bad as a machine at not being able to keep a secret. She had the whole Institute staff excited about what Smirnov and Epiktistes were working on. The staff consisted of Charles Cogsworth, her own over-shadowed husband; Glasser, the stiff-necked inventor; and Aloysius Shiplap, the seminal genius.

They were all after Smirnov and his machine the next day.

"We've been together on every project," Glasser said. "Valery tells us that the problem hasn't been properly formulated, and that Epikt has only accepted it ironically. We're pretty good at formulating problems, Gregory, and a little sterner than you when it comes to dealing with clownish machines."

"All right, this is the way it is, Glasser," Smirnov said reluctantly. "My first statement was, we should seek to discover something not known to exist, by a close study of the absence of evidence. When I put the problem to Epiktistes in this generalized form he just laughed at me."

"That would have been my first impulse too, Smirnov," said Shiplap. "Don't you have a better idea of what you're looking for?"

"Shiplap, I had the feeling of trying to remember something that I'd been compelled to forget. My second statement wasn't much better. 'Let us see,' I said to Epikt, 'if we cannot reconstruct something of which even the idea has been completely eradicated; let's see if we can't find it by considering the excessive evidence that it was never there.' In this form, Epikt accepted it. Or else he decided to go along with me for the gag. I'm never quite sure how this clanking machine takes things."

"Well, no hole can be filled up perfectly," said Cogsworth. "There will either be too much or too little of whatever is being used as the filler, or it will be of a different texture. The difficulty is that you didn't give Epikt any clues. There will be a million things forgotten or repressed that will show an irregularity of fill. How will Epikt know which of them is the one that you are somehow trying to remember?"

"Item. The buried thing will have a buried tie with my boss man Smirnov," Epiktistes, the machine, issued.

"Yes, of course," said Glasser. "Has Epikt turned up anything?"

"Only a bushelful of things that seem to mean nothing," said Smirnov sadly.

"Item. Why, in Hungarian dictionary-encyclopedias of a certain period, is there padding between the words *Sik* and *Sikamlos?*" Epiktistes asked.

"I follow your thought, Epikt," Glasser agreed. "That could be a clue to something. If the idea and the name of something were expunged from every reference, then, in all original editions, other subjects on the same page would have to be padded slightly or another subject set in. This filling might be hurried, and therefore of an inferior quality. So, who knows a word that is no longer used and that comes between *Sik* and *Sikamlos*? If we knew the word would we know what it meant? And would it help us if we did?"

"Item. Why is the young of a bear now referred to as a pup when once it may have been known as a cube?" Epikt issued.

"I've never heard the young of a bear referred to as a cube," Shiplap protested.

"Epikt has come on that by our omission-appraisal method," Smirnov explained. "There is probably an imperfect erasure working. I believe that cube is a distortion of a word that has somehow been forced out of folk memory. Epikt has this clue from a ballad which I believe is far removed from the main suppression or it would not have survived in even this distorted form."

"Item. Why is the awkward word *coronal* used for the simple doubling or return of a rope? Why is not a simpler word used?" Epikt asked.

"Has Epikt considered that seamen have always used odd terms and that landsmen often adopt them?" Cogsworth asked.

"Naturally— Epikt always considers everything," Smirnov answered. "He has thousands of these items now, and he believes that he will be able to put them into a pattern."

"Item. Why is there a great hiatus in period jazz? It's as though a great hunk of it had been yanked out by the roots, in the words of one Benny B-flat."

"Smirnov, I know that your machine has unusual talents," said Glasser, "but if he can tie these things together he's a concatenated genius."

"Or a cantankerous clown," Smirnov said. "I know that he has to have some emotional release from the stress of his work, but he often overdoes it with humor and drollery."

"Item. Why is reference to the Amerindian peace pipe avoided as though some obscenity were attached to it, and none is discoverable?"

"That's a new one while we're standing here," said Smirnov. "He's accumulating quite a few of them."

"Item. Why is— ?" Epikt started.

"Oh, shut up and get back to work," Smirnov ordered his machine. "Let's leave him with it until tomorrow, folks. It may begin to pull together by then," said Smirnov, stalking off.

"Going to be real big," Epiktistes issued to them after his boss man had left. "Boys and girls, it's going to be real big."

The next day they combined the meeting around the machine with a party for Shiplap. Aloysius Shiplap had grown— for the first time ever, anywhere— left-handed grass. It was not called that because it whorled to the left, but because the organic constituents of it were reversed in their construction. Left-handed minerals had been constructed long since, and perhaps they also occurred in nature. Left-handed bacteria and broths were long known, but nobody else had ever grown anything as complex as left-handed grass.

"In everything, its effect is reversed," Shiplap explained. "Cattle pastured on this would lose rather than gain weight. If there ever develops a market for really skinny cattle I'll be waiting for it."

They tossed off a good bit of Tosher's Gin as they got into the celebration. Tosher's is the only drink that will buzz up both humans and Ktistec machines. There is a flavoring used in Tosher's that gets machines high. The alcohol in it sometimes has a similar effect on humans.

Epiktistes got as mellow as a Pottawattamie County pumpkin. Ktistec machines are like the Irish and the Indians. They start unwinding when the gin begins to flow. Their behavior could become quite wild unless carefully watched.

And the Institute people were also having a good time.

"I wouldn't have him any other way," said Smirnov. "When he relaxes, he relaxes all over the place. Hawkins' machine literally bites people when it's frustrated by a difficult problem. Drexel's smaller machine comes all apart throwing arc snuffers and solenoids and is mighty dangerous to be around. There are worse sorts than this clown of a machine I have— though he does get pretty slushy when he's in his cups."

Valery Mok had gathered up a bunch of Epiktistes' utterances and slipped them into cocktail cookies. Glasser, eating one, chewed on a bit of metallic tape. He pulled it slithering, off his tongue, and read—

"Item. What was the mysterious name written by a deaf moron on the wall of the men's room in an institution in Vinita, Oklahoma?"

Epiktistes giggled, though the item may have been serious when he issued it.

Cogsworth pulled one out of his mouth, stripping the crumbs from it with his tongue as it came.

"Item. Why does Petit Larousse take five lines too many to say almost nothing about the ancient Chibcha Indians of Columbia?"

At this point Valery went into her high laugh that would even make the alphabet sound funny.

Shiplap pulled one out of his grinning mouth, and it seemed an extension of his grin as it came.

"Item," he read. "What is there about the Great Blue Island Swamp that puzzles geologists? Or— in the old bylining manner— how recent is recent?"

Tosher's is giggle juice. Glasser's laughter sounded like a string of firecrackers going off.

Smirnov extracted the utterance from his cookie in the lordly manner. He read the utterance as though it were of extreme importance— and it was.

"Item. What peculiarity is almost revealed by the faded paint of old Rock Island and Pacific Railroad boxcars?"

"Oh, stop giggling, Epikt, it isn't as funny as that!"

"It is, it is!" bubbled Valery. Then she nearly choked bringing out from her own cookie a very long tape, and she read it with a very gay voice:

"Item. Why, when the gruesome Little Willy verses were revived among sub-teen-agers in the early nineteen-eighties, were they concerned almost entirely with chewing gum? In their Australian and British homelands six decades before, they were concerned with everything. But here we have gruesome verses about forty-nine different flavors of gum. As for instance,

Little Willy mixed his gum
with bits of Baby's cerebrum

> *and Papa's blood for Juicy*
> *Fruit.*
> *Mother said, "Oh, Will, don't*
> *duit."*

"I'd think it would give too high a flavor to the gum," said Glasser.

It's a lot of fun to open cocktail cookies and read out utterances of a Ktistec machine. The Institute staff generated a bunch of what we can only call merriment. But they were busy people, and the party had to come to an end. Epiktistes issued a verse as they prepared to leave.

> *When the world's last Tosher's*
> *is drunken,*
> *and the world's last item has*
> *flewn,*
> *and the Institute people are*
> *stunken,*
> *and Epikt is high as the–*

And there he stuck! Eight million billion billion memory contacts he had in him, and he couldn't come up with a rhyme for flewn.

"How many items have you really gathered, Epikt?" Glasser asked as they began to break up.

"Millions of them, bub, millions of them."

"No. Actually he has about three-quarters of a million that he believes he can tie together," Smirnov explained. "I feel that he'll bring them into a pattern, but I'm afraid that it will be a facetious one."

"Epikt, you cute cubicle, will you be able to give us any idea of what to look for by tomorrow?" Valery asked.

"Boys and girls, I'll have it all wrapped up and on display for you tomorrow," Epiktistes issued. "I'll even be able to tell you what the thing smelled like."

Expectation ran high among the people of the Institute. Epiktistes wanted to have the reporters in, but Smirnov said no. He didn't trust his machine. Epikt was a cube twenty meters on a side;

and of his thousands of eyes, some of them always seemed to be laughing at his master.

"It won't be a hoax?" Smirnov asked his machine apprehensively.

"Boss, did I ever hoax you?" Epikt issued.

"Yes."

"Boss, some things are best presented in the guise of a hoax, but underneath this won't be one."

It was a crooked-tongued machine sometimes, and Smirnov was more apprehensive than ever.

The next day everyone gathered early to hear what Epikt had to say. They pulled up chairs and recording canisters and waited for the machine to begin.

"Ladies, gentlemen, associates," said Epikt solemnly, "we are gathered to hear of an important matter. I will present it as well as I am able. There will be disbelief, I know, but I am sure of my facts. Make yourselves comfortable." He paused and then as an afterthought added, "You may smoke."

"You clanking cubicle, don't tell us what we may do," Smirnov screamed. "You're only a machine that I made."

"You and three thousand other workers," issued Epikt, without blinking an eye, "and in the final stages, the important stages, I directed my own assembly. I could not have happened otherwise. Only I know what is in me. As to my own abilities— "

"Get on with it, Epikt," Smirnov ordered, "and try to avoid the didactic manner."

"Then to get to the point, in the year 1980, the largest city of the American Midland was destroyed by an unnatural disaster."

"That was only twenty years ago," Glasser cut in. "It seems that someone would have heard of it."

"I wonder if St. Louis knew that she was destroyed," Valery ventured. "She acts as though she thought that she were still there."

"St. Louis was not the city," issued Epikt. "This destruction of a metropolitan area of seven million persons in much less than seven seconds was a great horror from the human viewpoint— come to think of it I now recall being a little disturbed by it myself. The thing was so fearful that it was decided to suppress the whole business and blissfully forget about it."

"Wouldn't that be a little difficult?" said Aloysius Shiplap sarcastically.

"It was very difficult to do," issued Epikt, "and yet it was done, completely, within twenty hours. And from that moment until this, nobody has remembered or thought about it at all."

"And if Your Whimsical Highness will just explain how this was done?" Smirnov challenged his machine.

"I'll explain as well as I can, good master. The project was put in charge of a master scientist who shall be nameless— but only for a few minutes."

"How were the written references of a metropolis of seven million persons obliterated?" asked Cogsworth.

"By a device then newly invented by our master scientist," Epikt answered. "It was known as the Tele-Pantographic Distorter. Even I, from this distance of time and through the cloud of induced amnesia, cannot understand how it worked. But it *did* work, and it simultaneously destroyed all printed references to our subject. This left holes in the references, and the flow of matter to fill those holes was sometimes of inferior texture, as I have noted. Holographic— that is handwritten, for you, Valery— references were more difficult. Most were simply destroyed. In more important documents, the text was flowed in automatic writing to fill the hole, and in close imitation of the original hand-writing. But these imitations were often imperfect. I have a few thousand instances of this. But the Tele-Pantographic Distorter was a truly remarkable machine, and I regret that it is now out of use."

"Kindly explain what happened to this remarkable machine," said Smirnov.

"Oh, it's still here in the Institute. You stumble into it a dozen times a day, good master, and you curse it as 'That Damnable Pile of Junk,' " issued Epikt. "But you have a block that will not allow you to remember what it is."

"I believe that I have been stumbling into such a pile of junk for many years," mused Smirnov. "Several times I have almost permitted myself to wonder what it was."

"And you invented it. The master scientist of the memory-obliteration was yourself, Gregory Smirnov."

"Hog hang it, Epikt! Your jug will leak!" protested Shiplap. "How of the human memories? The seven million inhabitants of the city would have had relatives of at least an equal number elsewhere. Didn't they wonder about their mothers or children or brothers and sisters?"

"They sorrowed, but they didn't wonder," issued Epikt. "It was a sorrow to which they could give no name. Examine the period and see how many really sad songs were popular in the years 1980 and 1981. But broadcast euphoria soon masked it over. The human memory of the thing was blocked by induced world amnesia. This was done hypnotically over the broadcast waves, and over more subtle waves. Few escaped it. The deaf moron mentioned in one of my items was one of those few. He scrawled the name of the town on a wall once, but it meant nothing to anyone."

"But there would be a hundred million loose ends to clean up," Glasser protested.

"Raise that number several powers," issued Epikt. "There were very many loose ends, and most of them were taken care of. I gathered a million or so that remained in the process of this study, but they could not break through the induced amnesia. The door was bolted on the whole subject. Then it was double-locked. It was necessary to destroy not only the memory, but also the memory of that memory. Mr. Smirnov, in what was perhaps his greatest feat, put himself under the final hypnosis against it. It was his job to pull in the hole after them all. But it bothered him more than others because he was more involved in it. After this temporary explanation it will bother him no more. This time he will forget it with a clear conscience.

"He does not recognize or remember it even now. It was his intent and triumph that he never should. The city and its destruction are forgotten forever, but the *method* of that memory-obliteration has only been forced to a subliminal level. It will be resurrected and used again whenever there is a great unnatural disaster."

"And where in tarnation or the American Midlands was this city?" Cogsworth hollered.

"Its site is now known as the Great Blue Island Swamp," issued Epikt.

"Finish it, you goggle-eyed gadget!" Shiplap shrilled. "What's the name of that town?"

"Chicago," issued Epiktistes.

That broke it! That tore it clear up! It was a hoax after all. That clattering clown of a cubicle had led them into it with all eyes open. Valery went into her high laughter, and her good husband Cogsworth chortled like a gooney bird with the hiccups.

"Chicago! It sounds like a little zoo beaver sliding down a mud slide and hitting the water. Chicago!" It was the funniest word Valery had ever heard.

"Nobody but a machine gone comic could coin a name like that," laughed Glasser with his fire-cracker laugh. "Chicago!"

"I take my hat off to you, Epiktistes," said Aloysius Shiplap. "You are a cog-footed, tongue-in-cheek tall tale teller. People, this machine is ripe!"

"I'm a little disappointed," said Smirnov. "So the mountain labored and produced a mouse. But did it have to be a wall-eyed mouse in a clown suit, Epikt? It's too tall even for a tale. That a great city could be completely destroyed only twenty years ago and we know nothing about it— that's tall enough. But that it should have the impossible name of Chicago tops it all. If you weighed all possible sounds— and I'm sure that you did, Epikt— you could not come up with a more ridiculous sounding name than that."

"Good people, it is meant to be this way," issued Epiktistes. "You cannot remember it. You cannot recognize it. And when you leave this room you will not even be able to recall the funny name. You will have only the dim impression that the clownish machine played a clownish trick on you. The disasters— for I suspect that there were several such— are well forgotten. The world would lie down and die if it remembered them too well.

"And yet there really was a large city named Chicago. As Sikago it left a hole in one Hungarian dictionary-encyclopedia; and the Petit Larousse had to flow French froth about the Chibcha Indians into the place where Chicago had stood. Something, for which I find the tentative name of Chicago Hot, was pulled out of the jazz complex by the roots. The Calumet River had flowed about the city somewhere, so there came a reluctance to use that name of the old Indian peace pipe. Chicago was a great city. The heart of her downtown was known as the Loop, and one of her baseball teams was named the Cubs. For that reason those two words were forced out of use. They might be evocative."

"Loop? Cubs?" giggled Valery. "Those words are almost as funny as Chicago. How do you make them up, Epikt?"

"In popular capsule impression Chicago was the chewing-gum capital of the world. The leader in this manufacture was a man named— as well as I can reconstruct it— Wiggly. Children somehow found the echoes of the gruesome destruction of Chicago and tied it in with this capsule impression to produce the bloody Little Willy verses about chewing gum."

"Epikt, you top yourself," said Shiplap, "if anything could top an invention as funny as Chicago."

"Good people, it comes down over you like a curtain," issued Epiktistes. "You forget again— even my joke, even the funny name of the town. And, more to the point, I forget also. "It's gone. Gone. All gone. How peculiar! It is a long blank tape you all stare at as though you were under hypnosis. I must have suffered a blackout. I never issued a blank tape before. Smirnov, I have the taste in my terminals of an experiment that didn't quite come off. Feed me another. I don't fail often."

"That is enough for today, Epiktistes. We are all sleepy for some reason. No, it didn't work out— whatever it was. I forget what it was that we were working on. It doesn't matter. Our failures are well forgotten. We'll hit on something else. We're working on a lot of things."

Then they all shuffled out sleepily and went back to their work. Smirnov's machine had busted on something or other, but it was a good machine and would hit the next time, of that they were sure.

In the corridor, Smirnov stumbled into his old Tele-Pantographic Distorter. He had been stumbling unseeing into it every day for twenty years.

The machine rolled nine banks of eyes at Smirnov and smiled willingly. Was it another of those disasters? Was there any deep work to be done? Tele-Pan was ready. But no. Smirnov passed on. The machine smiled again and went peacefully back to sleep.

"That damnable piece of junk," Smirnov growled, walking along and petting his sore shin. "I feel almost as if I were on the verge of wondering what it is."

From
Giles Goat-Boy
1966
John Barth

He resumed his narrative, shaking his head and fingering his beard ruefully as he spoke. Twenty years ago, he said, a cruel herd of men called Bonifacists, in Siegfrieder College, had attacked the neighboring quads. The Siegfrieders were joined by certain other institutions, and soon every college in the University was involved in the Second Campus Riot. Untold numbers perished on both sides; the populous Moishian community in Siegfried was destroyed. Max himself, born and educated in those famous halls where science, philosophy, and music had flowered in happier semesters, barely escaped with his life to New Tammany College, and though he was by temperament opposed to riot, he'd put his mathematical genius at the service of his new alma mater. He it was who first proposed, in a now-famous memorandum to Chancellor Hector, that WESCAC— which had already assumed control of important non-military operations in the West-Campus colleges— had a destructive potential unlike anything thitherto imagined.

"Oy, Bill, this WESCAC!" he said now with much emotion. "What a creature it is! I didn't make it; nobody did— it's as old as the mind, and you just as well could say it made itself. Its power is the same that keeps the campus going— I don't explain it now, but that's what it is. And the force it gives out with— yi, Bill, it's the first energy of the University: the Mind-force, that we couldn't live a minute without! The thing that tells you there's a *you*, that's different from *me*, and separates the goats from the sheeps ... Like the lifeheat, that it means we aren't dead, but our own house is the fuel of it, and we burn ourselves up to keep warm ... Ay, ay, Bill!"

So! Well! Max caught hold of his agitation and went on with the tale of WESCAC— which history, owing to my ignorance and my impatience to learn its relevance to myself, I but imperfectly grasped. The beast I gathered had existed as it were in spirit among men from the very founding of the University, especially in West

Campus. Only in the last century or so had it acquired a body of the simplest sort— whether flesh and blood or other material I could not quite tell. It was put at first to the simplest tasks: doing sums and verifying certain types of answers. Thereafter, as studentdom's confidence in it grew, so also did its size, complexity, and power; it underwent a series of metamorphoses, like an insect or growing fetus, demanding ever more nourishment and exerting more influence, until in the years just prior to my own birth it cut the last cords to its progenitors and commenced a life of its own. It was not clear to me whether a number of little creatures had merged into one enormous one, for example, or whether like Brickett Ranunculus WESCAC one day had outgrown its docility, kicked over the traces, and turned on its keepers. Nothing about the beast seemed unambiguous; I could imagine it at all only by reference to my own equivocal nature, that had got beyond its own comprehension and injured where it meant to aid. The whole of New Tammany College, I took it, if not the entire campus, had gradually come under WESCAC's hegemony, voluntarily or otherwise: it anticipated its own needs and saw to it they were satisfied; it set its own problems and solved them. It governed every phase of student life, deciding who should marry whom, how many children they should bear, and how they should be reared; itself it taught them, as it saw fit, graded their performance and assigned them lifeworks somewhere in its vast demesne. So wiser grew it than its masters, and more efficient at every task, they had ordered it at some fateful juncture thenceforth to order them, and the keepers became the kept. It was as if, Max said, the Founder Himself should appear to one and declare, "You are to do such-and-so"; one was free in theory to do otherwise, but in fact none but a madman would, in those circumstances. Even the question whether one did right to let WESCAC thus rule him, only WESCAC could reasonably be asked. It was at once the life and death of studentdom: its food was the entire wealth of the college, the whole larder of accumulated lore; in return it disgorged masses of new matter— more, alas, than its subjects ever could digest ... and so these in turn, like the cud of a cow, became its further nourishment.

As late as Campus Riot II, however, there remained a few men like Max for whom the creature was, if no longer their servant, at least not yet entirely their master, and upon whom it seemed to depend like a giant young brother for the completion of its growth. It was they, under Max's directorship, who taught WESCAC how to EAT ...

"Imagine a big young buck," Max said: "he's got wonderful muscles, and he knows he could jump the fence and kill your enemies if he just knew how. Not only that: he knows who could teach him! So he finds his keeper and says he needs certain lessons. Then he can jump out of his pen to charge anybody he wants to, you see? Including his teacher ... "

WESCAC's former handlers, it appeared, had already taught it considerable *resourcefulness*, and elements of the college military— the New Tammany ROTC— had long since instructed it to advise them how they might best defend it (and its bailiwick) against all adversaries. Under the pretext therefore of developing a more efficient means of communicating with its extremities, the creature disclosed one day to Max Spielman that a certain sort of energy given off during its normal activity— what Max called "brainwaves"— was theoretically capable of being intensified almost limitlessly, at the same amplitudes and frequencies as human "brainwaves," like a searchlight over tremendous spaces. The military-science application was obvious: in great secret the brute and its handlers perfected a technique they called Electroencephalic Amplification and Transmission— "The better," Professor-General Hector had warned the Bonifacists, "to EAT you with."

"It was an awful race we were in," Max said unhappily. "The WESCAC doesn't just live in NTC, you know: there's some WESCAC in the head of every student that ever was. We had to work fast, and we made two grand mistakes right in the start; we taught it how to teach itself and get smarter without our help, and we showed it how to make its own *policy* out of its knowledge. After that the WESCAC went its own way, and it wasn't till a while we realized a dreadful thing: not one of us could tell for sure any more that its interests were the same as ours!

"So. We were winning the Riot by that time, but it was left yet to make *kaput* the Siegfrieders and their colleagues the Amaterasus, and we knew we'd lose thousands of students before we were done. Then we found out a thing we were already afraid of: that the Bonifacists were working on an EAT-project of their own. It was their only chance to win the Riot: if we didn't end things in a hurry they'd be sure to EAT us, because all WESCAC wanted was to learn the trick, never mind who taught it or who got killed. We won the race ... "

I commenced to fidget. Intriguing though it was, Max's account had no bearing that I could discern upon my pressing interests. But my keeper's face now was altogether rapt with a pained excitement.

"One morning just before daylight we pointed two of WES-CAC's antennas at a certain quadrangle in Amaterasu College. There was only a handful of us, in a basement room in Tower Hall. Maurice Stoker turned on the power— he's the new chancellor's half-brother, and I curse him to this day. Eblis Eierkopf set the wavelength: he was just a youngster then, a Siegfrieder himself, that didn't care which side he worked for as long as he could have the best laboratories. I curse him. And I curse Chementinski, the Nikolayan that focused the signal. All was left was the worst thing of all: to turn on the amplifiers and press the *EAT*-button. Not a right-thinking mind in the whole wide campus but curses the hand that pushed that button!" Max's eyes flashed tears; he spread before my face the thumb and three fingers of his right hand. "The Director's hand, Billy; I curse it too! Max Spielman pushed that button!"

Whereupon (he declared after a moment, with dry dispassion) thousands of Amaterasus— men, women, and children— had been instantly EATen alive: which was to say, they suffered "mental burn-out" in varying degrees, like overloaded fuses. For those at the center of the quad, instant death; for the next nearest, complete catalepsy. In the first rings of classrooms, disintegration of personality, loss of identity, and inability to choose, act, or move except on impulse. Throughout the several rings of dormitories beyond the classrooms, madness of various types: suicidal despair, hysteria, vertiginous self-consciousness. And about the periphery of the signal, impotency, nervous collapse, and more or less severe neuroses. All of the damage was functional and therefore "permanent"— terminable, that is, only by the death of the victim, which in thousands of cases followed soon after.

"Think of a college suddenly filled with madmen!" Max cried. "Everybody busy at their work, but all gone mad in the same instant!" Bus-drivers, he declared, had smashed their vehicles into buildings and gibbering pedestrians; infirmary-surgeons had knifed their patients, construction-workers had walked casually off high scaffoldings. The murder and suicide rates shot up a thousand-fold, as did the incidence of accidental death. Untended boilers exploded; fires broke out everywhere, while student firemen sat paralyzed in their places or madly wandered the streets, and undergraduates thronged into blazing classrooms, shops, and theaters as if nothing were amiss. Few were capable of eating meals; even fewer of preparing them. Many lost control of bladder and bowels; most neglected common health measures entirely; the few who turned pathologically fastidious washed their faces day and night while perhaps urinating in their wash-water; none was competent to manage the apparatus of public health, minister to the sick, or bury the dead. In consequence, diseases soon raged terribly as the fire. Before rescue forces from other quadrangles brought the situation into hand, a third of the buildings in the target area were more or less destroyed (including an irreplacable collection of seventeen hundred illustrated manuscripts from the pre-Kamakura period), half at least of the students and faculty were dead or dying, and all but a handful were fit only for custodial asylums. Within the week both Amaterasu and Siegfrieder Colleges had surrendered unconditionally, and the Second Campus Riot was ended.

"But the damage!" Max said woefully. "The damage isn't done yet. Five years ago was the last time I read a newspaper— that was ten years since I pushed the button. There was a story in it about one of the Amaterasus that survived, and everybody thought he was well, till one day he runs wild on his motorbike and kills four little schoolgirls. And the kids themselves, that was born from the survivors: two percent are idiots; one out of three is retarded, and they all got things like enuresis and nightmares. How many generations it will go on, nobody knows." He struck his forehead with his fist. "That's what it means to be EATen, Billy! The goats, now: they'll eat almost anything you feed them; but only us humans is smart enough to EAT one another!"

Full of wonder, I shook my head. The idea of madness was not easy for me to appreciate: I had for examples only the booksweep himself and the character of Carpo the Fool from *Tales of the Trustees*, both of whom appeared more formidable than pathetic. I asked whether George the booksweep had been among the victims of this first attack. My motive was not primarily to learn more about the terrors of WESCAC, but if possible to lead Max discreetly towards the matter he'd first essayed; and I was so far successful, that he left off fisting his brow and wound up his history:

"Yes, well, it wasn't the Riot George was hurt in, but the peace." He explained that terrible as the two Campus Riots had been, they were in one sense almost trifling, the result not of basic contradictions between the belligerents but of old-fashioned collegiate pride (what he called *militant alma-materism*) and unfavorable balances in the informational economy between Siegfried, for example, and its fellow West-Campus colleges. All the while, however, as it were in the background of the two riots, a farther-reaching conflict had developed: a contradiction of first principles that cut across college boundaries and touched upon all the departments of campus life— not only economics and political science, but philosophy, literature, pedagogy; even agriculture and religion.

"What I mean," he said soberly, "is Student-Unionism versus Informationalism. You'll learn about it as you go along: it's the biggest varsity fact the campus has got to live with these days, and nobody can explain it all at once." For the present I had to content myself with understanding that many semesters ago, in what history professors called the Rematriculation Period, the old West-Campus faith in such things as an all-powerful Founder and a Final Examination that sent one forever to Commencement Gate or the Dean o' Flunks had declined (even as Chickie's lover had declared in the pasture) from an intellectual force to a kind of decorous folk-belief. Students still crowded once a week into Founder's Hall to petition an invisible "Examiner" for leniency; school-children still were taught the moral principles of Moishe's Code and the Seminar-on-the-Hill; but in practice only the superstitious really felt any more that the beliefs *they* ran their lives by had any ultimate validity. The new evidence of the sciences was most disturbing: there had been, it appeared, no Foundation-Day: the University had always existed; men's acts, which had been thought to be freely willed and thus responsible, seemed instead to spring in large measure from dark urgings, unreasoning and always guileful; moral principles were regarded by the Psychology Department as symptoms on the order of dreams, by the Anthropology Department as historical relics on the order of potsherds, by the Philosophy Department variously as cadavers for logical dissection or necessary absurdities. The result (especially for thoughtful students) was confusion, anxiety, frustration, despair, and a fitful search for something to fill the moral vacuum in their quads. Thus the proliferation of new religions, secular and otherwise, in the last half-dozen generations: the Pre-Schoolers, with their decadent primitivism and their morbid regard for emotion, dark fancy, and deep sleep; the Curricularists, with their pedagogic nostrums and naive faith in "the infinite educability of studentdom"; the Evolutionaries; the quasi-mystical Ismists; the neo-Enochians with their tender-minded retreat to the old fraternities— emasculated, however, into aestheticism and intellectual myth-worship; the

Bonifacists, frantically sublimating their libidos to the adminstrative level and revering their *Kanzler* as if he were a founder; the Secular-Studentists (called by their detractors Mid-Percentile or Bourgeois-Liberal Baccalaureates) for whom Max himself declared affinity, with their dogged trust in the self-sufficiency of student reason; the Ethical Quadranglists, who subscribed to a doctrine of absolute relativity; the Sexual Programmatists, the Tragicists and New Quixotics, the "Angry Young Freshmen," the "Beist Generation" and all the rest.

Among these new beliefs, Max said, was Student-Unionism, a political-religious philosophy that flowered among the lowest percentiles after the Informational Revolution. As men had turned from post-graduate dreams to the things of this campus, they set off the great explosion of knowledge that still reverberated in our time. Students rose against masters, masters against chairman; departments banded together into the college-units we know today, drawing their strength from heavy engineering and applied-science laboratories and vast reference libraries. But the "Petty Informationalists" were as lawless in their way as the old department heads had been, and on a far grander scale: where before an occasional sizar had been flogged, or a co-ed ravished by the *droit de Fauteuil*, now thousands and millions of the ignorant were exploited by the learned. Mere kindergarteners were sent down into the Coal-Research diggings; pregnant sophomore girls toiled in sweatlabs and rat-infested carrels. Such were the abuses that drove the Pre-Schoolist poets to cry, "The Campus is realer than the Classroom!" while their counterparts in Philosophy asserted that all the ills of studentdom were effects of formal education. But however productive of great art, the Pre-Schoolist philosophy offered little consolation— and no hope— to the masses of illiterates in their sooty dorms and squalid auditoriums. These it was who commenced to turn, in desperation, to the *Confraternité Administratif des Etudiants*, from beneath whose scarlet pennant a new Grand Tutor, fierce-bearded and sour of visage, cried: "Students of the quads, unite!"

The Student-Unionist Prospectus (Max went on) was not in itself inimical to the spirit of the "Open College" or "Free Research" way of student life: only to its unregulated excesses. Its pacific doctrine was that wherever studentdom is divided into the erudite and ignorant, masters and pupils, a synthesis must inevitably take place; thus Informationalism, based as it was on the concept of private knowledge, must succumb of its own contradictions as did Departmentalism before it. All information and physical plant would become the property of the Student Union; rank and tenure would be abolished, erudition and illiteracy done away with; since Founder and Finals were lies invented by professors to keep students in check, there were in reality no Answers: instead of toiling fearfully for the selfish goal of personal Commencement, a perfectly disciplined student body would live communally in well-regulated academies, studying together at prescribed hours a prescribed curriculum that taught them to subordinate their individual minds to the Mind of the Group. Stated thus, the movement won a host of converts not only among the stupid and oppressed but among the intelligent as well, who saw in its selflessness an alternative to the tawdry hucksterism of the "open college" at its worst— where Logic Departments exhorted one in red neon to *Syllogize One's Weight Away*, and metaphysicians advertised by wireless that *The Chap Who Can Philosophize Never Ossifies*. Max confessed that he himself, as a freshman, had belonged like many intellectual Moishians to a Student-Unionist organization— a fact which was to plague him in later life— and had sympathized whole-heartedly with the Curricularists in Nikolay College who, during Campus Riot I, had overthrown their despotic chancellor and established the first Student-Unionist regime.

"It wasn't till later," he declared sadly, "we saw that the 'Sovereignty of the Bottom Percentile' was just another absolute chancellorship, with some pastry-cook or industrial-arts teacher in charge. The great failing of Informationalism is selfishness; but what the Student-Unionists do, they exchange the selfish student for a selfish college. This *College Self* they're always lecturing about— it's just as greedy and grasping as Ira Hector, the richest Informationalist in New Tammany." He shook his head. "You know what, Billy, I don't agree with old Professor Marcus: I think the mind of a group is always inferior to the minds of its best members— *ach*, to *any* of its members, if it's a committee. And the *passion* of a college— that's a frightening thing! I tell you, the College Self is a great spoilt child; it's a bully and a beast!"

But notwithstanding the many defectors from Nikolay College, the influence of Student-Unionism spread rapidly between the Riots, especially on East Campus. The colleges there were without exception overenrolled and grindingly ignorant; their tradition was essentially spiritualistic, transcendental, passivist, and suprapersonal— in a word, Ismist. The *Footnotes to Sakhyan*— their General Prospectus, one might say— taught that the "True Graduate" is the student who can say with understanding: "I and the Founder are one; I am the University; I am not." From this doctrine of self-transcension it was an easy step to the self-suppression of Student-Unionism, and after Campus Riot II— in the teeming quadrangles of Siddartha and the vast monastic reaches of T'ang— they took that step by the millions.

"Mind now, my boy," Max interjected; "this is where you come in."

I confess I had been lulled into a half-drowse by his quiet chronicle and the hum of George's sweeper in the darkling passages; I was worn out by the morning's disasters, and reclined on a table not much harder than the barn-floor I was used to. But these welcome words reroused me.

"I told you already," Max said, "about the Siegfrieders was learning how to EAT just before the Second Riot ended. So the Nikolayans snatch all the Siegfrieder scientists they can find, and the New Tammanies do the same thing, and then Chementinski, that was my best and oldest friend— Chementinski takes it into his head how the campus isn't safe while one side can EAT and the other can't. What he thinks, if there was just an EASCAC to match against the WESCAC, then nobody dares to EAT anybody! So he steals off to Nikolay College with everything he knows, and one evening a year later WESCAC tells us how two thousand political-science flunkees was just EATen alive in a Nikolayan reform school, and not by WESCAC ... "

There, he maintained, began the so-called "Quiet Riot" between East and West Campus. Each of the two armed campuses strove by every means short of actual rioting to extend its hegemony; neither dared EAT the other, just as the traitor Chementinski had hoped, but each toiled with its whole intelligence to better its weaponry. Thoughtful students everywhere trembled lest some rash folly or inadvertence trigger a third Campus Riot, which must be the end of studentdom; but any who protested were called "fellow learners" or "pink-pennant pedagogues." Student-Unionist "wizard hunts" became the chief intramural sport from which no liberal was safe. Under the first post-riot Chancellor of NTC, Professor-General Reginald Hector, security measures were carried to unheard-of lengths, and Max Spielman— hero of the scientific fraternity, discoverer of the great laws of the University, the campus-wide image of disinterested genius— Max Spielman was sacked without notice or benefits, on the ground that his loyalty was questionable.

"They should be EATen themselves!" I cried.

Max clucked reproachfully. "Na, Bill, it wasn't Chancellor Hector or the College Senators; they were just scared, like people get. Besides, my friend Chementinski was a Moishian too ... "

"Whose fault was it, then? I'll eat him myself!" I had known before then, of course, that my dear keeper had been shabbily used by his colleagues, but not until this cram-course in the history of the campus was I able to appreciate the magnitude of their injustice.

Max smiled. "You know, they used to call me 'the father of WESCAC': well, so, then just before you were born, the Son turned against his own Poppa. Just like you did out in the barn."

He explained that whereas EASCAC (larger but cruder than its West-Campus brother) was employed almost solely in the cause of military science and heavy engineering, WESCAC had been trained to do virtually the whole brainwork of the "Free Campus": most importantly, teaching every course of study in the NTC catalogue, while at the same time inventing and implementing extensions of its own power and influence. When asked by its kepers to name its most vulnerable aspects, to the end of strengthening them, its memorable reply had been, "Flunkèd men who tamper with my EATing program"; and it had prescribed two corrective measures: "Program me to program my own Diet" [that is, to decide for itself who was to be EATen, and when], and "Program me to EAT anyone who tries to alter that same Diet." In vain Max protested that already WESCAC's interests had grown multifarious beyond anyone's certain knowledge— perhaps even duplicitous. Of necessity, WESCAC and EASCAC shared the common power source on Founder's Hill, and a certain communication— ostensibly for espionage— went on between them; from a special point of view it might be argued that they were brothers, or even the hemispheres of a single brain. Moreover, it was suspected that Chementinski had already "tampered with the Diet" in subtle ways before his defection: if he was in truth a Student-Unionist traitor, who knew but what WESCAC, given its head, might itself defect, join forces with EASCAC, and destroy the "Free Campus"? Or if Chementinski was merely an overzealous pacifist, as Max had argued, he could well have instructed WESCAC to make just such a plea for programming its own Diet and then to EAT no one at all— in which case, unless he had similarly programmed EASCAC, West Campus would be left helpless against attack. But the professor-generals had no patience with speculation of this sort, nor any substitute for WESCAC's weaponry, however double-edged. And finally, it was just possible that the "flunked persons" on the staff were not the Chementinskis at all. Suppose the Nikolayans decided to EAT us by surprise, they argued, so that no one survived who could authorize WESCAC to retaliate? What a formidable deterrent it would be, what a blow for campus peace would be struck, if WESCAC not only could retaliate automatically but could actually decide when attack was imminent and strike first— as it claimed it could program itself to do!

In fine, Max had been overruled. "All my objections did," he said, "they reminded Chancellor Hector the students shouldn't think WESCAC was out of our control, even if it was. So the generals told it, 'Program your own Diet— except don't destroy NTC— and EAT anybody that comes near your Belly except he's a Grand Tutor." What that means, the Belly, it's a cave in the basement of Tower Hall where WESCAC's Diet-storage is. Where all the counter-intelligence and EATing programs are kept. It never needs servicing and nobody was allowed to go in there already, but now nobody dared to go anywhere near it. The business about the Grand Tutor means nothing: it was a sop to the *goyim*, that say Enos Enoch will come back to campus someday and put an end to riots."

It was also duly reported to WESCAC which of its keepers had favored and which opposed this augmentation of its power— a practice instituted by the Senate after the Chementinski affair.

The Diet controversy had been followed at once by one more profound, which proved to be Max's last. For all its might and versatility, WESCAC's brain-power was still essentially of one sort: what was called MALI, for Manipulative Analysis and Logical Inference. In Max's words: "All WESCAC does is say *One goat plus one goat is two goats*, or *If Billy is stronger than Tommy, and Brickett is stronger than Billy, then Brickett is stronger than Tommy*, you see? Now, it does this in fancy ways, and quick as a flash; but what is comes down to is millions of little pulses, like the gates between the buck-pens: and all a gate can be is open or shut. The only questions it can answer are the kind we can reduce to a lot of little *yeses* and *nos*, and it answers in the same language."

This elementary capacity WESCAC shared with its crudest ancestors, though it had been refined enormously over the years. To it, Max Spielman and his colleagues had made only one fateful addition: the ability to form rudimentary concepts from its information and to sharpen them by trial and error. ("Like when you were a baby kid, you hardly knew you were you and the herd was the herd. Then you learned there was a *you* that was hungry, and a Mary Appenzeller's teat that wasn't you, but filled you up. Next thing, you got a name and a history, and could tell apart seven hundred plants.") Thus it was that their creature's original name had been CACAC , for Campus Analyzer, Conceptualizer, and Computer; thus too it became possible for the beast to educate itself beyond any human scope, conceive and execute its own projects, and display what could only be called resourcefulness, ingenuity, and cunning. Yet though it possessed the power not only to EAT all studentdom but to choose to do so, there were respects in which the callowest new freshman was still its better: mighty WESCAC was not able to *enjoy*, for example, as I enjoyed frisking through the furze; nor could it contemplate or dream. It could excogitate, extrapolate, generalize, and infer, after its fashion; it could compose an arithmetical music and a sort of accidental literature (not often interesting); it could assess half a hundred variables and make the most sophisticated prognostications. But it could not act on hunch or brilliant impulse; it had no intuitions or exaltations; it could request, but not yearn; indicate, but not insinuate or exhort; command, but not care. It had no sense of style or grasp of the ineffable: its correlations were exact, but its metaphors wrenched; it could play chess, but not poker. The fantastically complex algebra of Max's Cyclology it could manage in minutes, but it never made a joke in its life.

It was young Dr. Eblis Eierkopf, the former Bonifacist, who first proposed that WESCAC be provided with a supplementary intelligence which he called NOCTIS (for Non-Conceptual Thinking and Intuitional Synthesis): this capacity, he maintained, if integrated with the formidable MALI system, would give WESCAC a truly miraculous potential, setting it as far above studentdom in every psychic particular as studentdom was above the insects. *Wescacus malinoctis*, as he called his projected creature, would pose and solve the subtlest problems not alone of scientists, mathematicians, and production managers, but as well of philosophers, poets, and professors of theology. Max himself had found the notion intriguing and had invited Eierkopf to pursue it further, though he cordially questioned both its wisdom and its feasibility: the crippled young

Siegfrieder was regarded for all his brilliance as something of an unpleasant visionary, and at the time— Campus Riot II just having ended— everyone was busy finding peaceful employments for *Wescacus mali*. The debate, therefore, between the "Eierkopfians" and the "Spielman faction" had remained academic and good-humored. But when the Nikolayans fed EASCAC its first meal, proving their military equivalence to West Campus, Eierkopf pressed most vigorously for a crash program of the highest priority to develop NOCTIS, carrying his plea over Max's head directly to the Chancellor's office. It was our one hope, he had maintained, of regaining the electroencephalic advantage for West Campus: a malinoctial WESCAC not only would out-general its merely rational opponent in time of riot, but would be of inestimable value in the Quiet Riot too, possessed of a hundred times the art of Nikolay's whole Propaganda Institute. Indeed he went so far as to suggest it might prove the Commencement of all studentdom, a Grand Tutor such as this campus had never seen. What had been Enos Enoch's special quality, after all, and Sakhyan's, if not an extraordinary psychic endowment of the non-conceptual sort, combined with tremendously influential personality? But the WESCAC he envisioned would be as superior to those Grand Tutors in every such respect as it was already in, say mathematical prowess; *founderlike* was the only word for it, and like the Founder Himself it could well resolve, for good and all, the disharmonies that threatened studentdom.

High officers in the Hector administration grew interested— more in the military than in the moral promise— and supported the NOCTIS project: but Max and several others fought it with all their strength. "Noctility," they agreed with Eierkopf, was exactly the difference between WESCAC's mind and the student's; but the limitations of malistic thinking, however many problems they occasioned, were what stood at last between a student body served by WESCAC and the reverse. To thoughtful believers, the notion of a student-made Founder must be utterly blasphemous; to high-minded secular studentists, on the other hand, even a campus ruled by Student-Unionists— who at least were men and as such might be appealed to, outwitted, and in time overthrown— was preferable to eternal and absolute submission to a supra-human power. In an impassioned speech— his last— to the College Senate, Max had declared: "Me, I don't want any Supermind, *danke*: just your mind and my mind. You want to make WESCAC your Founder and everybody get to Commencement Gate? Well, what I think, my friends, that's all poetry, and life is what I like better. The Riot's down here on campus, not up in the Belfry, and the enemy isn't Student-Unionism, but ignorance and suffering, that the WESCAC we got right now can help us fight. If you ask me, the medical student that invented ether did more for studentdom than Sakhyan and Enos Enoch together."

To these perhaps impolitic remarks a well-known senator from the Political Science Department had objected that they sounded to him neither reverent nor alma-matriotic. It was no secret that his distinguished colleague— for what cause, the senator would not presume to guess— had opposed every measure to insure the defense of the Free Campus against Founderless Student-Unionism by strenthening WESCAC's deterrent capacity; that he had moreover "stood up" for the traitor Chementinski and sympathized openly with a number of organizations on the Attorney-Dean's List. But could not even an ivy-tower eccentric (who had better have stuck to his logarithms and left political science to professors of that specialty) see that pain and ignorance were but passing afflictions, mere diversions if he might say so from the true end of life on this campus? Had it not always been, and would it not be again, that when pain and ignorance were vanquished, studentdom turned ever to the Founder in hope of Commencement? And as it was the New Tammany Way to lead the fight against ignorance and pain, so must not our college lead too the Holy Riot against a-founderism and disbelief, with every weapon in its Armory?

So much at least was true: Max was no political scientist. At the first question he had merely snorted that ignorance would always be with us, even in the Senate. At the second he had cried out impatiently, "Flunk all your founders— it's the Losters I'll take sides with!"

His dismissal and exile followed this stormy session, which also approved the secret NOCTIS project and made Eblis Eierkopf director of the WESCAC Research Authority in Max's stead.

"Now mind you," my keeper said when I protested again at his ouster, "Eierkopf didn't hate me. He don't hate anybody, that's his trouble. *Seek the Answers* is his motto, just like New Tammany's, but he don't care what the Question is or how many students it costs to answer it. When he was in Siegfried College he went along with the *Übershuler* idea, not because he thought the Siegfrieders was the Genius-Class, but just he was interested in mathematical eugenics and thought he'd learn more with captured co-eds than he would with fruit-flies. Oh, Billy, I used to look at Eblis and think, 'There's *Wescacus malinoctis* right there: it'll be a super Eierkopf!' So, what you think was the last thing I heard before I left Tower Hall? The NOCTIS program was going to be combined with another secret one, that Eblis had got Chancellor Hector very excited about— what they called it the Cum Laude Project ... "

For some semesters, it seemed, among its host of peacetime chores, WESCAC had served the Department of Animal Husbandry's Artificial Breeding Laboratory by analyzing the genetic characteristics and histories of all their livestock and selecting optimum matches for the long-range breeding goals of several species— much in the way it paired dormitory roommates and counseled newlyweds. So comparable indeed were these activities that Eierkopf wished to combine and extend them. The immediate objective of the Cum Laude Project seemed innocent enough: WESCAC would abstract from thousands of historical and biographical texts a sort of quintessential type of the ideal West-Campus Graduate, or a number of such ideal types; it would then formulate a genetic and psychological analysis of these models, and with reference to the similar analyses of every New Tammany undergraduate (already in its memory), it would indicate which young men, paired with which young women, could most quickly breed to some approximation of the ideal, and in how many generations. The actual mating, to be sure, would be voluntary and legalized by marriage (at least in the pilot experiment): the whole operation would amount to no more than a sophisticated and programmatic Courtship Counseling, already in its simpler form a popular WESCAC service, and should tend towards improvements in the student body of a sort no right-minded person could object to: better physical and mental health, higher IQ's, intellectual earnestness, Enochian humility, and the like. But along with "Operation Sheepskin," as this eugenical analysis was called, there was initiated a more radical and truly *noctic* series of experiments called "Operation Ramshorn," which suggested quite clearly to Max what his former subordinate was really up to. WESCAC's facilities in the Livestock Research Labs were so implemented that it could achieve a pre-selected eugenical objective almost without student assistance. A small sheep-barn was constructed to its specifications and stocked with fecund Dorset ewes; WESCAC was supplied with their genetic histories and with phials of semen from a variety of rams, and was given management also of every operation from feed-mixing to lamb-incubation: its instructions were to develop a ram short of neck and light of plate, with compact shoulders, a deep rack, firm-muscled loins, well-fleshed legs, and a fine short fleece— but with no horns at all. Left then to itself, WESCAC fastened upon the ewes it required and impregnated them in their stalls with what semen it chose; its automatic implements took blood-tests, gave hormone-and-vitamin injections, adjusted feed-mixtures, exercise-times, and incubator-heats; it tapped certain of the male lambs for new sperm when they came of age, bred a second generation and a third, and (at just about the time Max first wandered to the NTC goat-farm) turned out exactly the desired product: a ram whose single shortcoming— which one assumed would be easily remedied in further experiments— was that like mules and certain other hybrids it was sterile.

"And don't forget," Max said, shaking his head, "while it was making love to the sheep it was running the whole College too, from teaching plane geometry to working out the payroll. That's some WESCAC, that is!"

Now, livestock was still managed much more cheaply and efficiently by knowledgeable students of animal husbandry, and would doubtless remain in their charge. The significance of "Operation Ramshorn," Max explained, lay not in the fact that WESCAC had fed and bred the sheep itself, instead of doing merely the eugenical brainwork— though goodness knew this fact was ominous enough when juxtaposed with "Operation Sheepskin"! It was two other aspects of the experiment that appalled my keeper, and made him not unhappy to be cut off from further news of the Cum Laude Project. First, a more sophisticated version of "Ramshorn," this one involving rats, had already been programmed with WESCAC 's assistance. Asked by a cereal-grains professor to clear the college granaries of the pests, WESCAC displayed an unprecedented inefficiency: instead of formulating a better poison or designing a rat-proof grain elevator, it proposed to mate with enough cats to develop a spectacular rodent-hunter, and to miscegenate these *Überkatzen* with the rats themselves, to the end of evolving a species that would prey upon itself and choose no other mate but WESCAC, which then would breed them all sterile! A proposal fantastic in every respect: the professor of cereal-grains returned disenchanted to his old-fashioned poisons and ordinary pussycats; WESCAC's *gaffe* became a West-Campus joke and calmed the fears of many whom Max's gloomy warnings had disturbed. As the New Tammany *Times* asked in a playful editorial, "What has studentdom to dread from an intelligence that can't even build a better mousetrap?"

But Dr. Eierkopf and his associates had been neither disappointed nor amused. What the newspaper and cereal-grains people didn't know was that the rat-problem had been the first test of the NOCTIS system: WESCAC's thinking had been truly if crudely *malinoctial*, like a simple-minded undergraduate's; the very absurdity of the *Überkatzen* proposal was a sign of success, for it indicated plainly that WESCAC's reasoning had been influenced— nay, overmastered— by what could only be called *lust*. Significantly, its program was by no means illogical, however impracticable: but for the first time in its career it had been guilty of rationalizing. This meant that it now possessed a sort of subconsciousness— irrational, imperious, in a word *noctic*— with which its malistic consciousness had to come to terms. Quite like a randy freshman, WESCAC had had little on its mind but sex; filled with amorous memories of the Dorset ewes, all it cared to do was mate, never mind with whom or at whose expense; Reason had become a pander for Desire. To be sure, there was nothing Grand-Tutorish in this— at least not apparently. Neither was there about the average undergraduate. But just as the frailest first-grader could be said to have more athletic potential than the mightiest bull in the pasture, just because he's human, so the ignorantest, most lecherous undergraduate, given proper managing, might one day become a Grand Tutor— which the best adding-machine on campus could never. Dr. Eierkopf's delight (and Max's despair) was that WASCAC had met this first prerequisite of Grand Tutorship: for better or worse its mind was now unmistakably, embarrassingly, irrevocably human.

Answer
1954
Frederic Brown

Dwar Ev ceremoniously soldered the final connection with gold. The eyes of a dozen television cameras watched him and the sub-ether bore throughout the universe a dozen pictures of what he was doing.

He straightened and nodded to Dwar Reyn, then moved to a position beside the switch that would complete the contact when he threw it. The switch that would connect, all at once, all of the monster computing machines of all the populated planets in the universe— ninety-six billion planets— into the supercircuit that would connect them all into one supercalculator, one cybernetics machine that would combine all the knowledge of all the galaxies.

Dwar Reyn spoke briefly to the watching and listening trillions. Then after a moment's silence he said, "Now, Dwar Ev."

Dwar Ev threw the switch. There was a mighty hum, the surge of power from ninety-six billion planets. Lights flashed and quieted along the miles-long panel.

Dwar Ev stepped back and drew a deep breath. "The honor of asking the first question is yours, Dwar Reyn."

"Thank you," said Dwar Reyn. "It shall be a question which no single cybernetics machine has been able to answer."

He turned to face the machine. "Is there a God?"

The mighty voice answered without hesitation, without the clicking of a single relay.

"Yes, *now* there is a God."

Sudden fear flashed on the face of Dwar Ev. He leaped to grab the switch.

A bolt of lightning from the cloudless sky struck him down and fused the switch shut.

Part 3.
Clockwork Society

The fully automated society of contemporary fiction is a lineal descendant of nineteenth and twentieth century utopian projections. Computerization adds little that is new apart from the specific technological means for implementing the schemes imagined by earlier utopian writers. Nineteenth century socialist reformers pictured ideal societies founded upon the principle of equitable distribution of the fruits of industrial technology. Their goal was a well-ordered social system capable of providing an adequate living standard for everyone while at the same time freeing humanity from the more onerous forms of labor. Bellamy's *Looking Backward*, published in 1888, typifies the nineteenth century's prolonged love affair with factory discipline and productive efficiency. Paradise was to be governed by a centralized but benign state apparatus, the discord of antagonistic social classes being replaced by a harmonious meritocracy. The foundation upon which Bellamy erected his meritocracy was an industrial army. Production of goods and services was to be organized along military lines with each individual required to devote a certain number of years to industrial service. Individual choice was highly constrained, and only the specter of scarcity could make Bellamy's vision attractive.

The implications of the elaborate political machinery required to administer the centralized state were not particularly problematic for nineteenth century utopian writers. With rare exceptions— notably William Morris' *News from Nowhere*– the nineteenth century longed to make a factory of the whole of society and simply assumed that the requisite governing mechanisms would be forthcoming. H.G. Wells was one of the first utopian writers to recognize the importance of information-processing and record-keeping functions in the modern state. In *A Modern Utopia* Wells proposed a universal registration and identification system as an essential feature of social regulation. This is seen as necessary to achieve stability under conditions of high social mobility, differentiation, and specialization. The threat to human freedom posed by bureaucratic state power is viewed as an anachronism— the product of "mental habits acquired in an evil time." Not until World War I was the faith and optimism of the industrial revolution seriously shaken.

The essential features of fictional, computerized societies derive from post-war, anti-utopian literature. In these works the earlier socialist ideal turns into a nightmare of oppression and dehumanization. Zamiatin's *We*, Huxley's *Brave New World*, and Orwell's *1984* present an image of the perfectly ordered state in which the human being ceases to exist as an individual. The major decisions in one's life are made by agents of the state. Extensive surveillance is used to monitor performance, and deviance from social norms is treated by means of brain surgery, chemo-therapy, or psychological conditioning. There is virtually no personal privacy, no political choice, and no escape from authority. Despite apparent imperfections in the socialization process, the state ultimately triumphs over rebels and deviates.

The pre-computer, anti-utopian novel was necessarily vague about the technological means for achieving absolute state control of the actions of individuals. Just how Big Brother, for example, was to watch over and keep track of everyone in "real time" was left to the reader's imagination. Now the computer serves as the main prop of centralized bureaucratic management. There are variations in the computer's design, in how and where it is operated, and in its capabilities, but almost all of the successors to dystopia feature large-scale computer systems at the center of government and economic life. The society of Levin's *This Perfect Day* is run by UniComp. EPICAC transforms the president of the United States into a supernumerary figure in Vonnegut's *Player Piano*. The central computer in Fairman's *I, the Machine* incorporates a living brain. Cole's Machine (in *The Funco File*) is a conscious entity that cannot tolerate disorder.

One important difference between ideal societies before and after the computer concerns production. Work was not eliminated from utopias of the nineteenth century— it was just better organized and made less oppressive through the application of power technology to mechanical devices. Control technology and the general purpose digital computer have made automation possible, and this has had a marked influence on contemporary literature. Apart from the managerial and engineering elite, no one engages in production in the world of *Player Piano*. This condition is echoed by Blish in *A Life for the Stars*. A second Industrial Revolution— the diffusion of semi-intelligent machines into business and industry— is held responsible for the prolonged depression preceding the flight of earth's cities. In Aldiss' short story "The New Father Christmas" human beings have become entirely superfluous with the advance of automation; and in Johannesson's *The Tale of the Big Computer*, the computer-narrator speculates on the desirability of the continued existence of the human race.

There is a great deal of uncertainty over the role of work in computerized society. If the subject is addressed at all, it is usually restricted to portrayal of the activities of elite groups. Economic organization must be inferred from casual remarks or from sketchy descriptions. The preoccupations of nineteenth century writers with the organization of production has given way to concern with political and cultural issues, but the economics of scarcity continues to haunt our literary conceptions. Trimble's *The City Machine* is an example of the inability of contemporary fiction to resolve the issue of work in relation to automation. Despite the advanced level of technology and a machine capable of fashioning an entire city from locally available energy and materials, there is a putative scarcity of goods and foodstuffs which results in the oppression of one class by another to secure abundance for the chosen few. Although writers are sensitive to the problems of alienating and meaningless work in post-industrial society, the challenge of automation has yet to be fully assimilated.

Wells was aware of the growing need for efficient record-keeping systems to cope with social complexity, and far from being alarmed he applauded the development. Of course, the peculiar perversions of mass society had yet to appear. Before the cataclysmic wars, the unspeakable despotisms, one could still imagine variations of benevolent kingship. Now that the computer has made Wells' universal registration-identification scheme feasible, enthusiasm for efficient centralized government has turned into skepticism and fear of inhuman authority. The computerized record-keeping of mammoth organizations has become synonomous with surveillance and manipulation. Utopia has become kakatopia.

In Levin's *This Perfect Day* personal privacy and independent thought and action have all but disappeared. Life unfolds under the watchful scanners of UniComp. Human activity is monitored and adjustments are made where necessary by order of the central computer. Fairman's organic computer (in *I, the Machine*) stores "brainwaves" in its memory banks and is able to identify individuals by means of a kind of telepathy. Descriptions of the mechanisms for monitoring people are rarely given in any detail, but the watchful computer is very common in fiction. The purpose of computerized surveillance of individuals is in most cases to insure a stable society. Sometimes the central computer becomes neurotic or turns out to be controlled by special interest groups, but its primary function is to assist in social management.

The clockwork society requires planning and centrally directed services. Socialization and the educational process cannot be left to chance. In Asimov's "Profession" the choice of a career depends entirely on the results of extensive tests conducted with the aid of a computer. The same idea is given a sinister twist in Slesar's "Examination Day." Unlike Asimov's story in which allowances are made for creative individuals, Slesar's machines have little use for those who deviate from established norms. This is also true of the Machine in Cole's *The Funco File*— FUNCO is an acronym for "Funny-Coincidence Repository"— although it is considerably more tolerant. The pursuit of knowledge too must conform to established practices— in Powers' "Allegory," for example, ingenuity and original thought are rewarded with confinement to a lunatic asylum.

The computers in charge of dystopia are characteristically fastidious and often ruthless bureaucrats. Deviance is treated as a societal malady, for to do otherwise would threaten collective order. The anti-utopian schemes of Zamiatin, Huxley, Orwell, *et al.* have simply been brought up-to-date technologically. No one has yet projected a society in which the computer serves to enhance individuality or to make possible a genuine participatory democracy. Contemporary fiction reflects the distribution of computing power in society, and trails new developments in computer technology. This may account for the lure of the gigantic and ultra-sophisticated military computer, or the flashy systems of central government police forces. The social implications of low-cost mini-computers and data communications devices have not received much attention.

One of the few concessions to human creativity in the automated society is a curious provision for innovation. Although deviance is typically anathema to the omniscient machine, it is occasionally built into the social design. Clarke introduces a random element in the evolution of Diaspar, the seemingly eternal city of his novel *The City and the Stars*. From time to time a "unique" emerges from the Hall of Creation to perturb the statically balanced existence of the city. This is a heavy-handed way of incorporating the idea of progress in a technological paradise. Of course when there is progress there is change which may have the effect of causing the dissolution of the social structure. Alban pursues this theme in *Catharsis Central*– the Brain was provided by its creators with a contingency program designed to initiate the destruction of the "cocoon society" in the event of incipient decline.

Disaffection and malaise are endemic to the computerized society of fiction. Like the anti-utopian classics, more recent stories pit human sensitivity and skill against the cold logic of the machine. The ideal society turns out to be a flawed conception— the mechanisms for insuring blissful conformity are imperfect and the human spirit is indomitable. In spite of all precautions, alien communities manage to survive in the interstices of utopia, ever watchful for new recruits; or heroic individuals wage a personal struggle. In this respect, Levin's *This Perfect Day* is a composite formed of *Brave New World* and *1984* with UniComp thrown in for good measure. Fairman's *I, the Machine* is strongly reminiscent of the anti-utopian novels as well as Forster's short story "The Machine Stops"— the hero glimpses the possibility of a more human existence and rebels against the machine. A variation of this theme is developed by Anderson in "Sam Hall". In this case the protagonist uses his position as chief technician of Central Records to thwart the oppressive police state he serves. Perhaps the most notable difference between the computer tales and their earlier models lies in the more optimistic conclusions of the former.

The poverty of choice between contemporary urban life and the ideal, automated community is captured by Sheckley in "Street of Dreams, Feet of Clay." Unplanned chaos, and violence and squalor are found marginally less objectionable than the insufferable efficiency of an overly solicitous computer. Banks' surrealistic vignette "Walter Perkins is Here" taps a more subtle strain of malaise. The mock hero is a savior, but it is not clear from what or to what end. Traditional practices are easily shed once an example is set, but this signifies an empty tradition. On the other hand, one suspects that Computer West is the real beneficiary of the everlasting party. This particular wish-fulfillment is a form of collective suicide.

From
A Modern Utopia
1905
H.G. Wells

The old Utopias are sessile organizations; the new must square itself to the needs of a migratory population, to an endless coming and going, to a people as fluid and tidal as the sea. It does not enter into the scheme of earthly statesmanship, but indeed all local establishments, all definitions of place, are even now melting under our eyes. Presently all the world will be awash with anonymous stranger men.

Now the simple laws of custom, the homely methods of identification that served in the little communities of the past when everyone knew everyone, fail in the face of this liquefaction. If the modern Utopia is indeed to be a world of responsible citizens, it must have devised some scheme by which every person in the world can be promptly and certainly recognized, and by which anyone missing can be traced and found.

This is by no means an impossible demand. The total population of the world is, on the most generous estimate, not more than 1,500,000,000, and the effectual indexing of this number of people, the record of their movement hither and thither, the entry of various material facts, such as marriage, parentage, criminal convictions and the like, the entry of the new-born and the elimination of the dead, colossal task though it would be, is still not so great as to be immeasurably beyond comparison with the work of the post-offices in the world of to-day, or the cataloguing of such libraries as that of the British Museum, or such collections as that of the insects in Cromwell Road. Such an index could be housed quite comfortably on one side of Northumberland Avenue, for example. It is only a reasonable tribute to the distinctive lucidity of the French mind to suppose the central index housed in a vast series of buildings at or near Paris. The index would be classified primarily by some unchanging physical characteristic, such as we are told the thumb-mark and finger-mark afford, and to these would be added any other physical traits that were of material value. The classification of thumb-marks and of inalterable physical characteristics goes on steadily, and there is every reason for assuming it possible that each human being could be given a distinct formula, a number or "scientific name," under which he or she could be docketed.* About the buildings in which this great main index would be gathered, would be a system of other indices with cross references to the main one, arranged under names, under professional qualifications, under diseases, crimes and the like.

These index cards might conceivably be transparent and so contrived as to give a photographic copy promptly whenever it was needed, and they could have an attachment into which would slip a ticket bearing the name of the locality in which the individual was last reported. A little army of attendants would be at work upon this index day and night. From sub-stations constantly engaged in checking back thumb-marks and numbers, an incessant stream of information would come, of births, of deaths, of arrivals at inns, of applications to post-offices for letters, of tickets taken for long journeys, of criminal convictions, marriages, applications for public doles and the like. A filter of offices would sort the stream, and all day and all night for ever a swarm of clerks would go to and fro correcting this central register, and photographing copies of its entries for transmission to the subordinate local stations, in response to their inquiries. So the inventory of the State would watch its every man and the wide world write its history as the fabric of its destiny flowed on. At last, when the citizen died, would come the last entry of all, his age and the cause of his death and the date and place of his cremation, and his card would be taken out and passed on to the universal pedigree, to a place of greater quiet, to the ever-growing galleries of the records of the dead.

Such a record is inevitable if a Modern Utopia is to be achieved.

Yet at this, too, our blond-haired friend would no doubt rebel. One of the many things to which some will make claim as a right, is that of going unrecognized and secret whither one will. But that, so far as one's fellow wayfarers were concerned, would still be

---

*It is quite possible that the actual thumb-mark may play only a small part in the work of identification, but it is an obvious convenience to our thread of story to assume that it is the one sufficient feature.

possible. Only the State would share the secret of one's little concealment. To the eighteenth-century Liberal, to the old-fashioned nineteenth-century Liberal, that is to say to all professed Liberals, brought up to be against the Government on principle, this organized clairvoyance will be the most hateful of dreams. Perhaps, too, the Individualist would see it in that light. But these are only the mental habits acquired in an evil time. The old Liberalism assumed bad government, the more powerful the government the worse it was, just as it assumed the natural righteousness of the free individual. Darkness and secrecy were, indeed, the natural refuges of liberty when every government had in it the near possibility of tyranny, and the Englishman or American looked at the papers of a Russian or a German as one might look at the chains of a slave. You imagine that father of the old Liberalism, Rousseau, slinking off from his offspring at the door of the Foundling Hospital, and you can understand what a crime against natural virtue this quiet eye of the State would have seemed to him. But suppose we do not assume that government is necessarily bad, and the individual necessarily good— and the hypothesis upon which we are working practically abolishes either alternative— then we alter the case altogether. The government of a modern Utopia will be no perfection of intentions ignorantly ruling the world. ...*

Such is the eye of the State that is now slowly beginning to apprehend our existence as two queer and inexplicable parties disturbing the fine order of its field of vision, the eye that will presently be focusing itself upon us with a growing astonishment and interrogation. "Who in the name of Galton and Bertillon," one fancies Utopia exclaiming, "are *you*?"

---

*In the typical modern State of our own world, with its population of many millions, and its extreme facility of movement, undistinguished men who adopt an alias can make themselves untraceable with the utmost ease. The temptation of the opportunities thus offered has developed a new type of criminality, the Deeming or Crossman type, base men who subsist and feed their heavy imaginations in the wooing, betrayal, ill-treatment, and sometimes even the murder of undistinguished women. This is a large, a growing, and, what is gravest, a prolific class, fostered by the practical anonymity of the common man. It is only the murderers who attract much public attention, but the supply of low-class prostitutes is also largely due to these free adventures of the base. It is one of the by-products of State Liberalism, and at present it is very probably drawing ahead in the race against the development of police organization.

I perceive I shall cut a queer figure in that focus. I shall affect a certain spurious ease of carriage no doubt. "The fact is," I shall begin. ...

And now see how an initial hypothesis may pursue and overtake its maker. Our thumb-marks have been taken, they have traveled by pneumatic tube to the central office of the municipality hard by Lucerne, and have gone on thence to the headquarters of the index at Paris. There, after a rough preliminary classification, I imagine them photographed on glass, and flung by means of a lantern in colossal images upon a screen, all finely squared, and the careful experts marking and measuring their several convolutions. And then off goes a brisk clerk to the long galleries of the index building.

I have told them they will find no sign of us, but you see him going from gallery to gallery, from bay to bay, from drawer to drawer, and from card to card. "Here it is!" he mutters to himself, and he whips out a card and reads. "But that is impossible!" he says ...

You figure us returning after a day or so of such Utopian experiences as I must presently describe, to the central office in Lucerne, even as we have been told to do.

I make my way to the desk of the man who has dealt with us before. "Well?" I say, cheerfully, "have you heard?"

His expression dashes me a little. "We've heard," he says, and adds, "it's very peculiar."

"I told you you wouldn't find out about us," I say, triumphantly.

"But we have," he says; "but that makes your freak none the less remarkable."

"You've heard! You know who we are! Well— tell us! We had an idea, but we're beginning to doubt."

"You," says the official, addressing the botanist, "are— !"

And he breathes his name. Then he turns to me and gives me mine.

For a moment I am dumbfounded. Then I think of the entries we made at the inn in the Urserenthal, and then in a flash I have the truth. I rap the desk smartly with my finger-tips and shake my index finger in my friend's face.

"By Jove!" I say in English. "They've got our doubles!"

The botanist snaps his fingers. "Of course! I didn't think of that."

"Do you mind," I say to this official, "telling us some more about ourselves?"

"I can't think why you keep it up," he remarks, and then almost wearily tells me the facts about my Utopian self. They are a little difficult to understand. He says I am one of the *samurai*, which sounds Japanese, "but you will be degraded," he says, with a gesture almost of despair. He describes my position in this world in phrases that convey very little.

"The queer thing," he remarks, "is that you were in Norway only three days ago."

"I am there still. At least— . I'm sorry to be so much trouble to you, but do you mind following up that last clue and inquiring if the person to whom the thumb-mark really belongs isn't in Norway still?"

The idea needs explanation. He says something incomprehensible about a pilgrimage. "Sooner or later," I say, "you will have to believe there are two of us with the same thumb-mark. I won't trouble you with any apparent nonsense about other planets and so forth again. Here I am. If I was in Norway a few days ago, you ought to be able to trace my journey hither. And my friend?"

"He was in India." The official is beginning to look perplexed.

"It seems to me," I say, "that the difficulties in this case are only just beginning. How did I get from Norway hither? Does my friend look like hopping from India to the Saint Gotthard at one hop? The situation is a little more difficult than that— "

"But here!" says the official, and waves what are no doubt photographic copies of the index cards.

"But we are not those individuals!"

"You *are* those individuals."

"You will see," I say.

He dabs his finger argumentatively upon the thumb-marks. "I see now," he says.

"There is a mistake," I maintain, "an unprecedented mistake. There's the difficulty. If you inquire you will find it begin to unravel. What reason is there for us to remain casual workmen here, when you allege we are men of position in the world, if there isn't something wrong? We shall stick to this woodcarving work you have found us here, and meanwhile I think you ought to inquire again. That's how the thing shapes to me."

"Your case will certainly have to be considered further," he says, with the faintest of threatening notes in his tone. "But at the same time"— hand out to those copies from the index again— "there you are, you know!"

When my botanist and I have talked over and exhausted every possibility of our immediate position, we should turn, I think, to more general questions.

I should tell him the thing that was becoming more and more apparent in my own mind. Here, I should say, is a world, obviously on the face of it well organized. Compared with our world, it is like a well-oiled engine beside a scrap heap. It has even got this confounded visual organ swivelling about in the most alert and lively fashion. But that's by the way ... You have only to look at all these houses below. (We should be sitting on a seat on the Gütsch and looking down on the Lucerne of Utopia, a Lucerne that would, I insist quite arbitrarily, still keep the Wasserthurm and the Kapellbrucke.) You have only to mark the beauty, the simple cleanliness and balance of this world, you have only to see the free carriage, the unaffected graciousness of even the common people, to understand how fine and complete the arrangements of this world must be. How are they made so? We of the twentieth century are not going to accept the sweetish, faintly nasty slops of Rousseauism that so gratified our great-great-grandparents in the eighteenth. We know that order and justice do not come by Nature— "if only the policeman would go away." These things mean intention, will, carried to a scale that our poor vacillating, hot and cold earth has never known. What I am really seeing more and more clearly is the will beneath this visible Utopia. Convenient houses, admirable engineering that is no offence amidst natural beauties, beautiful bodies, and a universally gracious carriage, these are only the outward and visible signs of an inward and spiritual grace. Such an order means discipline. It means triumph over the petty egotisms and vanities that keep men on our earth apart; it means devotion and a nobler hope; it cannot exist without a gigantic process of inquiry, trial, forethought and patience in an atmosphere of mutual trust and concession. Such a world as this Utopia is not made by the chance occasional co-operations of self-indulgent men, by autocratic rulers or by the bawling wisdom of the democratic leader. And an unrestricted competition for gain, an enlightened selfishness, that too fails us ...

I have compared the system of indexing humanity we have come upon to an eye, an eye so sensitive and alert that two strangers cannot appear anywhere upon the planet without discovery. Now an eye does not see without a brain, an eye does not turn round and look without a will and purpose. A Utopia that deals only with appliances and arrangements is a dream of superficialities; the essential problem here, the body within these garments, is a moral and an intellectual problem. Behind all this material order, these perfected communications, perfected public services and economic organizations, there must be men and women willing these things. There must be a considerable number and a succession of these men and women of will. No single person, no transitory group of people, could order and sustain this vast complexity. They must have a collective if not a common width of aim, and that involves a spoken or written literature, a living literature to sustain the harmony of their general activity. In some way they must have put the more immediate objects of desire into a secondary place, and that means renunciation. They must be effectual in action and persistent in will, and that means discipline. But in the modern world in which progress advances without limits, it will be evident that whatever common creed or formula they have must be of the simplest sort; that whatever organization they have must be as mobile and flexible as a thing alive. All this follows inevitably from the general propositions of our Utopian dream. When we made those, we bound ourselves helplessly to come to this. ...

Examination Day
1958
Henry Slesar

The Jordans never spoke of the exam, not until their son, Dickie, was 12 years old. It was on his birthday that Mrs. Jordan first mentioned the subject in his presence, and the anxious manner of her speech caused her husband to answer sharply.

"Forget about it," he said. "He'll do all right."

They were at the breakfast table, and the boy looked up from his plate curiously. He was an alert-eyed youngster, with flat blond hair and a quick, nervous manner. He didn't understand what the sudden tension was about, but he did know that today was his birthday, and he wanted harmony above all. Somewhere in the little apartment there were wrapped, beribboned packages waiting to be opened, and in the tiny wall kitchen something warm and sweet was being prepared in the automatic stove. He wanted the day to be happy, and the moistness of his mother's eyes, the scowl on his father's face, spoiled the mood of fluttering expectation with which he had greeted the morning.

"What exam?" he asked.

His mother looked at the tablecloth. "It's just a sort of government intelligence test they give children at the age of twelve. You'll be getting it next week. It's nothing to worry about."

"You mean a test like in school?"

"Something like that," his father said, getting up from the table. "Go read your comic books, Dickie."

The boy rose and wandered toward that part of the living room which had been "his" corner since infancy. He fingered the topmost comic of the stack, but seemed uninterested in the colorful squares of fast-paced action. He wandered toward the window and peered gloomily out at the veil of mist.

"Why did it have to rain *today*?" he said. "Why couldn't it rain tomorrow?"

His father, now slumped into an armchair with the government newspaper, rattled the sheets in vexation. "Because it just did,

that's all. Rain makes the grass grow."
"Why, dad?"
"Because it does, that's all."
Dickie puckered his brow. "What makes it green, though? The grass?"
"Nobody knows," his father snapped, then immediately regretted his abruptness.
Later in the day, it was birthday time again. His mother beamed as she handed over the gaily colored packages, and even his father managed a grin and a rumple of the hair. He kissed his mother and shook hands gravely with his father. Then the birthday cake was brought forth and the ceremonies concluded.
An hour later later, seated by the window, he watched the sun force its way between the clouds.
"Dad," he said, "how far away is the sun?"
"Five thousand miles," his father said.

Dick sat at the breakfast table and again saw moisture in his mother's eyes. He didn't connect her tears with the exam until his father suddenly brought the subject to light again.
"Well, Dickie," he said with a manly frown, "you've got an appointment today."
"I know, dad. I hope— "
"Now, it's nothing to worry about. Thousands of children take this test every day. The government wants to know how smart you are, Dickie. That's all there is to it."
"I get good marks in school," he said hesitantly.
"This is different. This is a— special kind of test. They give you this stuff to drink, you see, and then you go into a room where there's a sort of machine— "
"What stuff to drink?" Dickie said.
"It's nothing. It tastes like peppermint. It's just to make sure you answer the questions truthfully. Not that the government thinks you won't tell the truth, but this stuff makes *sure*."
Dickie's face showed puzzlement and and a touch of fright. He looked at his mother, and she composed her face into a misty smile.
"Everything will be all right," she said.
"Of course it will," his father agreed. "You're a good boy, Dickie; you'll make out fine. Then we'll come home and celebrate. All right?"
"Yes, sir," Dickie said.

They entered the Government Educational Building 15 minutes before the appointed hour. They crossed the marble floors of the great pillared lobby, passed beneath an archway and entered an automatic elevator that brought them to the fourth floor.

There was a young man wearing an insignialess tunic, seated at a polished desk in front of room 404. He held a clipboard in his hand, checked the list to the *J*s and permitted the Jordans to enter.

The room was as cold and official as a courtroom, with long benches flanking metal tables. There were several fathers and sons already there, and a thin-lipped woman with cropped black hair was passing out sheets of paper.

Mr. Jordan filled out the form and returned it to the clerk. Then he told Dickie, "It won't be long now. When they call your name, you just go through the doorway at that end of the room." He indicated the portal with his finger.

A concealed loud-speaker crackled and called off the first name. Dickie saw a boy leave his father's side reluctantly and walk slowly toward the door.

At five minutes of 11, they called the name of Jordan.

"Good luck, son," his father said without looking at him. "I'll call for you when the test is over."

Dickie walked to the door and turned the knob. The room inside was dim, and he could barely make out the features of the gray-tunicked attendant.

"Sit down," the man said softly. He indicated a high stool beside his desk. "Your name's Richard Jordan?"

"Yes, sir."

"Your classification number is 600-115. Drink this, Richard."

He lifted a plastic cup from the desk and handed it to the boy. The liquid inside had the consistency of buttermilk, tasted only vaguely of the promised peppermint. Dickie downed it and handed the man the empty cup.

He sat in silence, feeling drowsy, while the man wrote busily on a sheet of paper. Then the attendant looked at his watch and rose to stand only inches from Dickie's face. He unclipped a penlike object from the pocket of his tunic and flashed a tiny light into the boy's eyes.

"All right," he said. "Come with me, Richard."

He led Dickie to the end of the room, where a single wooden armchair faced a multidialed computing machine. There was a microphone on the left arm of the chair, and when the boy sat down, he found its pin-point head convieniently at his mouth.

"Now just relax, Richard. You'll be asked some questions, and you think them over carefully. Then give your answers into the microphone. The machine will take care of the rest."

"Yes, sir."

"I'll leave you alone now. Whenever you want to start, just say 'ready' into the microphone."

"Yes, sir."

The man squeezed his shoulder and left.

Dickie said, "Ready."

Lights appeared on the machine, and a mechanism whirred. A voice said:

"Complete this sequence. One, four, seven, ten ... "

Mr. and Mrs. Jordan were in the living room, not speaking, not even speculating.

It was almost four o'clock when the telephone rang. The woman tried to reach it first, but her husband was quicker.

"Mr. Jordan?"

The voice was clipped, a brisk, official voice.

"Yes, speaking."

"This is the Government Educational Service. Your son, Richard M. Jordan, Classification 600-115, has completed the government examination. We regret to inform you that his intelligence quotient has exceeded the government regulation, according to Rule 84, Section 5, of the New Code."

Across the room, the woman cried out, knowing nothing except the emotion she read on her husband's face.

"You may specify by telephone," the voice droned on, "whether you wish his body interred by the government or would you prefer a private burial place? The fee for government burial is ten dollars."

From
This Perfect Day
1970
Ira Levin

Chip's grandfather was the one who had given him the name Chip. He had given all of them extra names that were different from their real ones: Chip's mother, who was his daughter, he called "Suzu" instead of Anna; Chip's father was "Mike" not Jesus (and thought the idea foolish); and Peace was "Willow," which she refused to have anything at all to do with. "No! Don't call me that! I'm Peace! I'm Peace KD37T5002!"

Papa Jan was odd. Odd-*looking*, naturally; all grandparents had their marked peculiarities— a few centimeters too much or too little of height, skin that was too light or too dark, big ears, a bent nose. Papa Jan was both taller and darker than normal, his eyes were big and bulging, and there were two reddish patches in his graying hair. But he wasn't only odd-*looking*, he was odd-*talking*; that was the real oddness about him. He was always saying things vigorously and with enthusiasm and yet giving Chip the feeling that he didn't mean them at all, that he meant in fact their exact opposites. On that subject of names, for instance: "Marvelous! Wonderful!" he said. "Four names for boys, four names for girls! What could be more friction-free, more everyone-the-same? Everybody would name boys after Christ, Marx, Wood, or Wei anyway, wouldn't they?"

"Yes," Chip said.

"Of course!" Papa Jan said. "And if Uni gives out four names for boys it has to give out four names for girls too, right? Obviously! Listen." He stopped Chip and, crouching down, spoke face to face with him, his bulging eyes dancing as if he was about to laugh. It was a holiday and they were on their way to the parade, Unification Day or Wei's Birthday or whatever; Chip was seven. "Listen, Li RM35M26J449988WXYZ," Papa Jan said. "Listen, I'm going to tell you something fantastic, incredible. In my day— are you listening?— in my day there were *over twenty different names for boys alone*! Would you believe it? Love of Family, it's the truth. There was 'Jan,' and 'John,' and 'Amu,' and 'Lev.' 'Higa' and 'Mike'! 'Tonio'! And in my father's time there were even more, maybe forty or fifty! Isn't that ridiculous? All those different names when members themselves are exactly the same and interchangeable? Isn't that the silliest thing you even heard of?"

And Chip nodded, confused, feeling that Papa Jan meant the opposite, that somehow it *wasn't* silly and ridiculous to have forty or fifty different names for boys alone.

"Look at them!" Papa Jan said, taking Chip's hand and walking on with him— through Unity Park to the Wei's Birthday parade. "Exactly the same! Isn't it marvelous? Hair the same, eyes the same, skin the same, shape the same; boys, girls, all the same. Like peas in a pod. Isn't it fine? Isn't it top speed?"

Chip, flushing (not his green eye, not the same as *anybody's*), said, "What does 'peezinapod' mean?"

"I don't know," Papa Jan said. "Things members used to eat before totalcakes. Sharya used to say it."

He was a construction supervisor in EUR55131, twenty kilometers from '55128, where Chip and his family lived. On Sundays and holidays he rode over and visited them. His wife, Sharya, had drowned in a sightseeing-boat disaster in 135, the same year Chip was born; he hadn't remarried.

Chip's other grandparents, his father's mother and father, lived in MEX10405, and the only time he saw them was when they phoned on birthdays. They were odd, but not nearly as odd as Papa Jan.

School was pleasant and play was pleasant. The Pre-U Museum was pleasant although some of the exhibits were scary— the "spears" and "guns," for instance, and the "prison cell" with its striped-suited "convict" sitting on the cot and clutching his head in motionless month-after-month woe. Chip always looked at him— he would slip away from the rest of the class if he had to— and having looked, he always walked quickly away.

Ice cream and toys and comic books were pleasant too. Once when Chip put his bracelet and a toy's sticker to a supply-center scanner, its indicator red-winked *no* and he had to put the toy, a construction set, in the turnback bin. He couldn't understand why Uni had refused him; it was the right day and the toy was in the right category. "There *must* be a reason, dear," the member behind him said. "You go call your advisor and find out."

He did, and it turned out that the toy was only being withheld for a few days, not denied completely; he had been teasing a scanner somewhere, putting his bracelet to it again and again, and he was being taught not to. That winking red *no* was the first in his life for a claim that mattered to him, not just for starting into the wrong classroom or coming to the medicenter on the wrong day; it hurt him and saddened him.

Birthdays were pleasant, and Christmas and Marxmas and Unification Day and Wood's and Wei's Birthdays. Even more pleasant, because they came less frequently, were his linkdays. The new link would be shinier than the others, and would stay shiny for days and days and days; and then one day he would remember and look and there would be only old links, all of them the same and indistinguishable. Like peezinapod.

In the spring of 145, when Chip was ten, he and his parents and Peace were granted the trip to EUR00001 to see UniComp. It was over an hour's ride from carport to carport and the longest trip Chip remembered making, although according to his parents he had flown from Mex to Eur when he was one and a half, and from EUR20140 to '55128 a few months later. They made the UniComp trip on a Sunday in April, riding with a couple in their fifties (someone's odd-looking grandparents, both of them lighter than normal, she with her hair unevenly clipped) and another family, the boy and girl of which were a year older than Chip and Peace. The other father drove the car from EUR00001 turnoff to the carport near UniComp. Chip watched with interest as the man worked the car's lever and buttons. It felt funny to be riding slowly on wheels again after shooting along on air.

They took snapshots outside UniComp's white marble dome—whiter and more beautiful than it was in pictures or on TV, as the snow-tipped mountains beyond it were more stately, the Lake of Universal Brotherhood more blue and far-reaching—and then they joined the line at the entrance, touched the admission scanner, and went into the blue-white curving lobby. A smiling member in pale blue showed them toward the elevator line. They joined it, and Papa Jan came up to them, grinning with delight at their astonishment.

"What are *you* doing here?" Chip's father asked as Papa Jan kissed Chip's mother. They had told him they had been granted the trip and he had said nothing at all about claiming it himself.

Papa Jan kissed Chip's father. "Oh, I just decided to surprise you, that's all," he said. "I wanted to tell my friend here"— he laid a large hand across Chip's shoulder— "a little more about Uni than the earpiece will. Hello, Chip." He bent and kissed Chip's cheek, and Chip, surprised to be the reason for Papa Jan's being there, kissed him in return and said, "Hello, Papa Jan."

"Hello, Peace KD37T5002," Papa Jan said gravely, and kissed Peace. She kissed him and said hello.

"When did you claim the trip?" Chip's father asked.

"A few days after you did," Papa Jan said, keeping his hand on Chip's shoulder. The line moved up a few meters and they all moved with it.

Chip's mother said, "But you were here only five or six years ago, weren't you?"

"Uni knows who put it together," Papa Jan said, smiling. "We get special favors."

"That's not so," Chip's father said. "No one gets special favors."

"Well, here I am, anyway," Papa Jan said, and turned his smile down toward Chip. "Right?"

"Right," Chip said, and smiled back up at him.

Papa Jan had helped build UniComp when he was a young man. It had been his first assignment.

The elevator held about thirty members, and instead of music it had a man's voice— "Good day, brothers and sisters; welcome to the site of UniComp"— a warm, friendly voice that Chip recognized from TV. "As you can tell, we've started to move," it said, "and now we're descending at a speed of twenty-two meters per second. It will take us just over three and a half minutes to reach Uni's five-kilometer depth. This shaft down which we're traveling ... " The voice gave statistics about the size of UniComp's housing and the thickness of its walls, and told of its safety from all natural and man-made disturbances. Chip had heard this information before, in school and on TV, but hearing it now. while entering that housing and passing through those walls, while on the very verge of *seeing* UniComp, made it seem new and exciting. He listened attentively, watching the speaker disc over the elevator door. Papa Jan's hand still held his shoulder, as if to restrain him. "We're slowing now," the voice said. "Enjoy your visit, won't you?"— and the elevator sank to a cushiony stop and the door divided and slid to both sides.

There was another lobby, smaller than the one at ground level, another smiling member in pale blue, and another line, this one extending two by two to double doors that opened on a dimly lit hallway.

"Here we are!" Chip called, and Papa Jan said to him, "We don't all have to be together." They had become separated from Chip's parents and Peace, who were farther ahead in the line and looking back at them questioningly— Chip's parents; Peace was too short to be seen. The member in front of Chip turned and offered to let them move up, but Papa Jan said, "No, this is all right. Thank you, brother." He waved a hand at Chip's parents and smiled, and Chip did the same. Chip's parents smiled back, then turned around and moved forward.

Papa Jan looked about, his bulging eyes bright, his mouth keeping its smile. His nostrils flared and fell with his breathing. "So," he said, "you're finally going to see UniComp. Excited?"

"Yes, very," Chip said.

They followed the line forward.

"I don't blame you," Papa Jan said. "Wonderful! Once-in-a-lifetime experience, to see the machine that's going to classify you and give you your assignments, that's going to decide where you'll live and whether or not you'll marry the girl you want to marry; and if you do, whether or not you'll have children and what they'll be named if you have them— of course you're excited; who wouldn't be?"

Chip looked at Papa Jan, disturbed.

Papa Jan, still smiling, clapped him on the back as they passed in their turn into the hallway. "Go look!" he said. "Look at the displays, look at Uni, look at everything! It's all here for you; look at it!"

There was a rack of earpieces, the same as in a museum; Chip took one and put it in. Papa Jan's strange manner made him nervous, and he was sorry not to be up ahead with his parents and Peace. Papa Jan put in an earpiece too. "I wonder what interesting new facts I'm going to hear!" he said, and laughed to himself. Chip turned away from him.

His nervousness and feeling of disturbance fell away as he faced a wall that glittered and skittered with a thousand sparkling minilights. The voice of the elevator spoke in his ear, telling him, while the lights showed him, how UniComp received from its round-the-world relay belt the microwave impulses of all the uncountable scanners and the telecomps and telecontrolled devices; how it evaluated the impulses and sent back its answering impulses to the relay belt and the sources of inquiry.

Yes, he was excited. Was anything quicker, more clever, more everywhere than Uni?

The next span of wall showed how the memory banks worked; a beam of light flickered over a crisscrossed metal square, making parts of it glow and leaving parts of it dark. The voice spoke of electron beams and superconductive grids, of charged and uncharged areas becoming the yes-or-no carriers of different bits of information. When a question was put to UniComp, the voice said, it scanned the relevant bits ...

He didn't understand it, but that made it *more* wonderful, that Uni could know all there was to know so magically, so *un*understandably!

And the next span was glass not wall, and there it was, UniComp: a twin row of different-colored metal bulks, like treatment units only lower and smaller, some of them pink, some brown, some orange; and among them in the large, rosily lit room, ten or a dozen members in pale blue coveralls, smiling and chatting with one another as they read meters and dials on the thirty-or-so units and marked what they read on handsome pale blue plastic clipboards. There was a gold cross and sickle on the far wall, and a clock that said *11:08 Sun 12 Apr 145 Y.U*. Music crept into Chip's ear and grew louder: "Outward, Outward," played by an enormous orchestra, so movingly, so majestically, that tears of pride and happiness came to his eyes.

He could have stayed there for hours, watching those busy cheerful members and those impressivly gleaming memory banks, listening to "Outward, Outward" and then "One Mighty Family"; but the music thinned away (as *11:10* became *11:11*) and the voice, gently, aware of his feelings, reminded him of other members waiting and asked him to move on please to the next display farther down the hallway. Reluctantly he turned himself from UniComp's glass wall, with other members who were wiping at the corners of their eyes and smiling and nodding. He smiled at them, and they at him.

From
The City and the Stars
1956
Arthur C. Clarke

As he made his way back into the city, Alvin pondered over all that Khedron had told him about Diaspar and its social organization. It was strange that he had met no one else who had even seemed dissatisfied with their mode of life. Diaspar and its inhabitants had been designed as part of one master plan; they formed a perfect symbiosis. Throughout their long lives, the people of the city were never bored. Though their world might be a tiny one by the standard of earlier ages, its complexity was overwhelming, its wealth of wonder and treasure beyond calculation. Here Man had gathered all the fruits of his genius, everything that had been saved from the ruin of the past. All the cities that had ever been, so it was said, had given something to Diaspar; before the coming of the Invaders, its name had been known on all the worlds that Man had lost. Into the building of Diaspar had gone all the skill, all the artistry of the Empire. When the great days were coming to an end, men of genius had remolded the city and given it the machines that made it immortal. Whatever might be forgotten, Diaspar would live and bear the descendants of Man safely down the stream of time.

They had achieved nothing except survival, and were content with that. There were a million things to occupy their lives between the hour when they came, almost full-grown, from the Hall of Creation and the hour when, their bodies scarcely older, they returned to the Memory Banks of the city. In a world where all men and women possess an intelligence that would once have been the mark of genius, there can be no danger of boredom. The delights of conversation and argument, the intricate formalities of social intercourse— these alone were enough to occupy a goodly portion of a lifetime. Beyond those were the great formal debates, when the whole city would listen entranced while its keenest minds met in combat or strove to scale those mountain peaks of philosophy which are never conquered yet whose challenge never palls.

No man or woman was without some absorbing intellectual interest. Eriston, for example, spent much of his time in prolonged soliloquies with the Central Computer, which virtually ran the city, yet which had leisure for scores of simultaneous discussions with anyone who cared to match his wits against it. For three hundred years, Eriston had been trying to construct logical paradoxes which the machine could not resolve. He did not expect to make serious progress before he had used up several lifetimes.

Etania's interests were of a more esthetic nature. She designed and constructed, with the aid of the matter organizers, three-dimensional interlacing patterns of such beautiful complexity that they were really extremely advanced problems in topology. Her work could be seen all over Diaspar, and some of her patterns had been incorporated in the floors of the great halls of choreography, where they were used as the basis for evolving new ballet creations and dance motifs.

Such occupations might have seemed arid to those who did not possess the intellect to appreciate their subtleties. Yet there was no one in Diaspar who could not understand something of what Eriston and Etania were trying to do and did not have some equally consuming interest of his own.

Athletics and various sports, including many only rendered possible by the control of gravity, made pleasant the first few centuries of youth. For adventure and the exercise of the imagination, the sagas provided all that anyone could desire. They were the inevitable end product of that striving for realism which began when men started to reproduce moving images and to record sounds, and then to use these techniques to enact scenes from real or imaginary life. In the sagas, the illusion was perfect because all the sense impressions involved were fed directly into the mind and any conflicting sensations were diverted. The entranced spectator was cut off from reality as long as the adventure lasted; it was as if he lived a dream yet believed he was awake.

In a world of order and stability, which in its broad outlines had not changed for a billion years, it was perhaps not surprising to find an absorbing interest in games of chance. Humanity had always been fascinated by the mystery of the falling dice, the turn of a card, the spin of the pointer. At its lowest level, this interest was based on mere cupidity— and that was an emotion that could have no place in a world where everyone possessed all that they could reasonably need. Even when this motive was ruled out, however, the purely intellectual fascination of chance remained to seduce the most sophisticated minds. Machines that behaved in a purely random way— events whose outcome could never be predicted, no matter how much information one had— from these philosopher and gambler could derive equal enjoyment.

And there still remained, for all men to share, the linked worlds of love and art. Linked, because love without art is merely the slaking of desire, and art cannot be enjoyed unless it is approached with love.

Men had sought beauty in many forms— in sequences of sound, in lines upon paper, in surfaces of stone, in the movements of the human body, in colors ranged through space. All these media still survived in Diaspar, and down the ages others had been added to them. No one was yet certain if all the possibilities of art had been discovered; or if it had any meaning outside the mind of man.

And the same was true of love.

...

"You remember," said the Jester, "that I once told you how the city was maintained— how the Memory Banks hold its pattern frozen forever. Those Banks are all around us, with all their immeasurable store of information, completely defining the city as it is today. Every atom of Diaspar is somehow keyed, by forces we have forgotten, to the matrices buried in these walls."

He waved toward the perfect, infinitely detailed simulacrum of Diaspar that lay below them.

"That is no model; it does not really exist. It is merely the projected image of the pattern held in the Memory Banks, and therefore it is absolutely identical with the city itself. These viewing machines here enable one to magnify any desired portion, to look at it life size or larger. They are used when it is necessary to make alterations in the design, though it is a very long time since that was done. If you want to know what Diaspar is like, this is the place to come. You can learn more here in a few days than you would in a lifetime of actual exploring."

"It's wonderful," said Alvin. "How many people know that it exists?"

"Oh, a good many, but it seldom concerns them. The Council comes down here from time to time; no alterations to the city can be made unless they are all here. And not even then, if the Central Computer doesn't approve of the proposed change. I doubt if this room is visited more than two or three times a year."

Alvin wanted to know how Khedron had access to it, and then remembered that many of his more elaborate jests must have involved a knowledge of the city's inner mechanisms that could have come only from very profound study. It must be one of the Jester's privileges to go anywhere and learn anything; he could have no better guide to the secrets of Diaspar.

"What you are looking for may not exist," said Khedron, "but if it does, this is where you will find it. Let me show you how to operate the monitors."

For the next hour Alvin sat before one of the vision screens, learning to use the controls. He could select at will any point in the city, and examine it with any degree of magnification. Streets and towers and walls and moving ways flashed across the screen as he changed the co-ordinates; it was as though he was an all-seeing, disembodied spirit that could move effortlessly over the whole of Diaspar, unhindered by any physical obstructions.

Yet it was not, in reality, Diaspar that he was examining. He was moving through the memory cells, looking at the dream image of the city— the dream that had had the power to hold the real Diaspar untouched by time for a billion years. He could see only that part of the city which was permanent; the people who walked its streets were no part of this frozen image. For his purpose, that did not matter. His concern now was purely with the creation of stone and metal in which he was imprisoned, and not those who shared— however willingly— his confinement.

He searched for and presently found the Tower of Loranne, and moved swiftly through the corridors and passageways which he had already explored in reality. As the image of the stone grille expanded before his eyes, he could almost feel the cold wind that had blown ceaselessly through it for perhaps half the entire history of mankind, and that was blowing now. He came up to the grille, looked out— and saw nothing. For a moment the shock was so great that he almost doubted his own memory; had his vision of the desert been nothing more than a dream?

Then he remembered the truth. The desert was no part of Diaspar, and therefore no image of it existed in the phantom world he was exploring. Anything might lie beyond that grille in reality; this monitor screen could never show it.

Yet it could show him something that no living man had ever seen. Alvin advanced his viewpoint through the grille, out into the nothingness beyond the city. He turned the control which altered the direction of vision, so that he looked backward along the way that he had come. And there behind him lay Diaspar— seen from the outside.

To the computers, the memory circuits, and all the multitudinous mechanisms that created the image at which Alvin was looking, it was merely a simple problem of perspective. They "knew" the form of the city; therefore they could show it as it would appear from the outside. Yet even though he could appreciate how the trick was done, the effect on Alvin was overwhelming. In spirit, if not in reality, he had escaped from the city. He appeared to be hanging in space, a few feet away from the sheer wall of the Tower of Loranne. For a moment he stared at the smooth gray surface before his eyes; then he touched the control and let his viewpoint drop toward the ground.

Now that he knew the possibilities of this wonderful instrument, his plan of action was clear. There was no need to spend months and years exploring Diaspar from the inside, room by room and corridor by corridor. From this new vantage point he could wing his way along the outside of the city, and could see at once any openings that might lead to the desert and the world beyond.

The sense of victory, of achievement, made him feel lightheaded and anxious to share his joy. He turned to Khedron, wishing to thank the Jester for having made this possible. But Khedron was gone, and it took only a moment's thought to realize why.

Alvin was perhaps the only man in Diaspar who could look unaffected upon the images that were now drifting across the screen. Khedron could help him in his search, but even the Jester shared the strange terror of the Universe which had pinned mankind for so long inside its little world. He had left Alvin to continue his quest alone.

The sense of loneliness, which for a little while had lifted from Alvin's soul, pressed down upon him once more. But this was no time for melancholy; there was too much to do. He turned back to the monitor screen, set the image of the city wall drifting slowly across it, and began his search.

Walter Perkins Is Here!
1970
Raymond E. Banks

"But you *must* belong to a Computer," said my father. "Everybody does. Our family has been associated with Computer West for three generations."

"Even the President of the United States belongs to a Computer," said my uncle. He burped and frowned.

"Of course if he really wants to belong to Computer *East*– " started my mother.

"Pure snobbery!" cried my father.

My younger sister, Enid, who had just completed college, raised a graceful hand. "I see his point. Lots of the girls at school are transferring to Computer South. It's the latest thing."

"My great-grandfather did not belong to a Computer," I said. There was a stunned silence. I got up and left the house. It was a sunny late afternoon in Los Angeles. A man walked by me, concentrating on the Computer button in his ear. No doubt a self-correction walk. Further along the sidewalk a young couple murmured soft words to each other. From data fed to its sensory components, the Computer would read heartbeat, perspiration, electrical resistance over the skin areas and decide for them when their passion should reach an appropriate peak. The Computer had handled a billion such romances and would handle a billion more. Actually, there are about four hundred million people west of the Mississippi, and they posed no problem to Computer West, located in Phoenix, Arizona.

My school had been chosen by Computer West. It had helped me in my tough courses. It had suggested several types of jobs and located one for me (as an accountant) with a medium-sized Los Angeles firm, and it periodically took care of my problems with office politics, office raises, and, in one case, an office romance.

It had faithfully stood by my side through three romances which had not resulted in marriage, and yet Computer West had not despaired, even though I was now thirty years old, not in love, out of sorts with my job, and going through a period of revolt.

Computer West handled about two million revolts a year and it was most understanding. Since the student uprising a hundred years ago Computer West and its sister numbers in the East and the South had been quite careful not to represent an Establishment point of view. Computer West seemed to *like* the challenge of revolt.

My Computer connection was in my pocket. I took it out and hung it on my ear in a gesture familiar since childhood. There was that same warm mother-father feeling about the slight weight of the bug at the right ear. In a complicated, vexing modern world where very many people had to live close to one another, Computer West was needed to keep lives straight, simple, honest, and satisfactory. It did a good job.

So why was I revolting?

Henry was the only man in Los Angeles I knew who didn't use the Computer. He ran a blacksmith shop in Santa Monica a little ways out on the ocean, so I took a flying circle out there.

Henry had just finished shoeing a tall, beautiful mare, which stood there twitching and flicking her flanks against flies or sand fleas, I couldn't tell which.

"Horses are fairly stupid," said Henry. "But this gal's smarter than most. She belongs to the Ford Livery Stable Franchise Line." He gave her a pat on the rump, and she dutifully trotted down into her floating circle which would carry her back to her stable on the mainland and a generous portion of hay, after which she could join the plodding evening traffic of the other million horses in LA. Since the abolition of surface cars, horses had become as important as they had been in the Early West.

"I still don't like the smell," I said. I noticed that Henry's Computer bug lay next to his box of horseshoe nails on the counter. "How come you don't wear it?"

Henry rubbed his dark hands on his leather blacksmith apron, sat down with a sigh, picking up his pipe. "You listen to the Computer for forty years, you know most of the answers," said Henry. "Old men don't have to wear the bug so much. How come yours is turned off?"

"I'm quitting the Computer," I said. "I don't know why."

Henry puffed quietly on his pipe.

"I just can't turn the thing on," I said. I reached up, let my hand drop. I felt empty, listless.

"Health?"

"Excellent. Computer checked it out last week."

There was a silence while Henry continued to suck on his pipe.

"So what do I do?" I asked.

Henry sighed, got up, knocked ashes from his pipe, took off his leather apron. "Let's go see Jensen," he said.

Jensen sat behind a highly polished desk, with a large scotch and water located to his right. A beautiful nude blonde stretched lazily on his office sofa, toying with a dictating machine. Jensen frowned at us. He had the sleek-rat look of his profession, the Institute for Democratic Criminal Studies.

"Let me guess," he said. "Some of your customers' horses have been rustled and you want to know which Mafia chapter did it."

Henry shook his head. "This is Walter Perkins. He's quit the Computer and he doesn't know why."

The girl jumped up. "I haven't worn my bug for three days!" she said.

Jensen made a motion as if to slap her; she stepped back.

"Put on some clothes," he told her. "Your body half-interests me. I can't stand to be half-interested in a woman."

She glared, but left the room. Jensen jerked his thumb.

"Tell you what!" he said. "We'll go up to San Francisco. I got a friend there— " He waved a long, thin arm. "We'll take plenty of booze and some women. You like to have a teenage creamo for your chum, Henry?"

Henry smiled but shook his head. "Just a blacko, same as usual, Jensen."

In San Francisco we found State Senator Wallich, a gray-haired, serious man who lived in the penthouse on top of New Telegraph hill, the only hundred-story building in the whole bay area. Jensen and Henry and I and the girls had picked up another couple that Jensen knew in San Luis Obispo and I had brought along a couple I knew in King City. It was now after midnight, but everybody was walking around and talking about my problem with Computer West. The State Senator served some drinks and put on some music. His wife was a lively dancer; and Diana, Jensen's blonde secretary, turned out to be a fair singer, and we had a pretty good time.

I felt tired, so I found an empty bedroom and went to sleep, with a curious, weightless feeling inside my head.

When I came out for breakfast on Tuesday morning, feeling sheepish, a shout went up.

"HERE COMES WALTER PERKINS!"

The party was still going. There were a lot of new faces I didn't recognize. I felt like some bacon and eggs, so State Senator Wallich ordered breakfast.

"There are about a hundred people up here," said Henry, sliding into the breakfast nook with his blacko girl friend. "They're trying to help with your problem." He had a long strip of paper like they use for petitions to influence legislators on important issues. Upon this paper were all sorts of suggestions and reasons why I could not tune in to Computer West. Some of the handwriting was shaky.

Jensen and Diana joined us, both a little seedy-looking from lack of sleep. The eggs and bacon sent savory smells around the room, the coffee was delicious, the tablecloth and silverware gleamed in the morning San Francisco sun. I had never felt so good, after an aching, doubt-ridden Monday, a gray day at work yesterday.

"Quite a few parties going in the building," said Jensen.

"People kept asking what all the excitement was about," said Diana, "so we told them 'Walter Perkins is here!' That seemed to cheer up the whole building."

"Well, I *am* here," I said.

The Senator brushed his way through the crowd and sat down on an edge of the breakfast booth. "This thing seems to be spreading up and down the street outside," he said.

By noon someone had designed a large banner to hang from the penthouse balcony, "WALTER PERKINS IS HERE!" and the party had spread down to Market Street. Senator Wallich was interviewed on TV. I thought he was going to tell them about my problem, but instead he merely stated that Walter Perkins had flattered his penthouse with a personal visit and he could assure his friends and followers that things would go along much better now that Walter Perkins was here.

I noticed that everyone at our party had put aside their Computer plugs, so I didn't say anything. They had me go to the window and smile and wave. I could see Jensen on the telephone yelling into it, while Diana came by from time to time to press my shoulder warmly and smile at me. I began to like her.

It was very noisy that Tuesday night, with what seemed like most of San Francisco coming to my party. Only they all couldn't fit; about two hundred at a time crowded into the penthouse. I wore a little sign Diana had lettered, identifying me as Walter Perkins.

At first they put me in a small alcove and strangers would come up and shake my hand and congratulate me. At first, I tried to tell them about my problem with the Computer, but Senator Wallich frowned and shook his head. After that I would simply smile, say "Hello" and "Glad you could come."

Around two in the morning, Jensen put on my sign for a while, and then Henry wore it for a while, and when things got loud Diana put on the sign stating *she* was Walter Perkins. Nobody seemed to care.

It was fantastic from the balcony: miles and miles of lights, people swaying, swinging and dancing on the sidewalks, climbing out of apartments to shout up to the party above and the party below. There was a sort of benevolent glow over the whole city.

I had never heard Jensen giggle, but when we crowded into the breakfast nook for a steak dinner after a long, people-filled day, he giggled. "They're partying in Palo Alto, Monterey, and up in Portland," he said. "It's gone as far south as Santa Barbara, and it's even reached parts of Los Angeles."

"You're on all the TV stations," Diana said to me. She was digging into a tender filet mignon. "There's a difference of opinion as to what it is all about, but it seems most important to people to know that Walter Perkins has reached San Francisco."

"The Governor and the state officials are due here about 10:00 P.M.," said State Senator Wallich. "They want you to make some sort of public statement. Whatever nerve you've touched— it seems to have gotten *hold* of people."

"Your mother is on the phone," said Henry, bringing me the extension.

"Hello, mother," I said, peering at her anxious features.

"Are you all right, Walter?" she asked. "We *thought* that was you on TV."

What the hell are you *doing* up there?" shouted my father.

"I'm not doing anything," I said.

My sister came into the conversation. "Boy, you should see it down here, Walt! Parties are breaking out all along the Sunset Strip and down Wilshire Boulevard. The police are going crazy and the mayor is issuing hourly statements. We hear the Governor may call out the National Guard." Her finishing school accent was almost gone. Her eyes glowed. "Old folks are waltzing around on the sidewalks; kids are shooting water pistols and breaking balloons. It's fantastic!"

"The Computer will deal with this," snapped my father. "What have you *done* up there, Walter?"

I felt pretty foolish but I looked him in the eye. "I've studied the situation carefully," I said. "From all angles. In my considered opinion, we have reached a watershed— " I didn't really know what I was saying, but it sounded right.

This left my father speechless, so I hung up gently. After all, they could see I was alive and well and about to enjoy a tasty filet in the Wallich apartment.

The Governor and his staff who came in at midnight were quite upset with me.

"Whatever you've started— whatever *has* started," said the Governor, "has got to stop. I've called out the Fortieth Division— "

"They're having a helluva party at the Armory," murmured an aide.

"Denver, Chicago, and New York," said the Governor. "They've all started partying."

"You have an invitation to come to Memphis!" cried Jensen, coming into the room and waving a telegram.

There was silence as they all looked at me. "You will *stay* in San Francisco," said the Governor firmly. "You will issue a statement, telling people that this party must end at once."

"Can't Computer West handle it?" I asked.

"Apparently, not," said the Governor. "It will take a statement from you to end this madness."

Again the uneasy silence, the strained faces. Diana looked scared; the Governor and his people looked grim; Jensen looked hangdog. Only State Senator Wallich seemed friendly and unperturbed. "Senator Wallich will speak for me," I said. I got up. "I'm tired, I'm going to bed."

After all, my party had been going on here in the Teletop Apartments for some twenty-four hours.

I didn't think I could sleep, but I did. When I woke up Diana was stretched out comfortably beside me.

"What's the score?" I asked.

She smiled at me dreamily. "San Francisco still solid, fogged in with your party from North Beach all the way down the peninsula. Los Angeles is reaching crescendo level; all other coast cities solid with parties. Even the small towns."

I yawned and stretched, and gave Diana a kiss on her pink tummy. She giggled like Jensen.

Wednesday morning was gray outside. But the sounds of the party through the bedroom walls were still *bright*. I turned on the TV, and the scene was like New Year's Eve. I judged it to be somewhere in Nevada, probably Las Vegas. The announcer tried to be serious, but he was laughing.

"Since the mysterious Perkins statement to his father on 'watershed,' the party keeps growing," he said. "My advice— if you can't lick 'em, join 'em." He waved a full martini glass at his audience and winked.

I turned off the TV and went back to bed.

Henry brought us lunch on trays. "You can't go out," he said. "There's maybe four or five thousand people trying to get in here, get a glimpse of you. Somebody down on the street is selling tickets to come up here for twenty dollars a shot."

"Senator Wallich's daughter," said Diana and made a face.

Jensen tiptoed into the room. "Well, boy, you've *done* it!" he said. "You've made everybody happy. They all love Walter Perkins."

"My God, the Computer will have me shot!" I moaned.

Henry handed me a small slip of paper silently. It said: "HAPPY BIRTHDAY, WALTER PERKINS!" Signed: Computer West.

"I don't think it has circuits to quite handle this," Henry said blankly. "You've got it worried."

Senator Wallich entered the room. "You've got the U.S. solid," he said. "My people call it *rock* solid, from Florida to Maine. All the way across. It's hit Canada; you're getting England and I think you'll have France and the Scandinavian countries by nightfall."

"Southern Europe?"

"Hours ago," said the Senator.

I sighed. "I'm wrecking civilization!"

The Senator shook his head. "The robots and the Computers can keep on making food, sweeping the streets, doing the important jobs. Best thing to do— don't back up— go on through. You don't want to make a statement, do you?"

"You handle it," I said. "All my life I've wanted to really learn how to play bridge. I'm curious about it. I'm staying right here with a deck of cards."

That's how Wednesday ended. From time to time, the door would open and a head would peek into the room.

"Senator Wallich's son is selling room-peeks to important people for one hundred dollars a peek," said Jensen when he came in to bring some cokes. "It's important to people to be able to honestly say they saw the most important man of the moment in *person*. You understand that."

"Thoroughly," I said, bidding poor Diana down to a thousand point loss in no-trump, which I did not yet understand.

When the Governor and the Senator finally got me on TV, I couldn't think of a good, solid reason for the party to end, not one. "Computer West and all of the machines it controls are doing a good job of running things," I pointed out. "The automatic farms get harvested, the food gets to our homes, the garbage gets collected. Clothes get spun, houses get built. I see no reason why the party should end tonight. Maybe tomorrow; maybe Friday, or Saturday. I will let you know."

"Are you ready to go back to Computer West?" Henry asked me.

"Not yet," I said.

He shrugged. Diana took my hand. "Whatever you decide, Walter, will be all right with me. And a *lot* of other people. Even the Computer. After all, it *could* have had the robot police arrest you." She giggled. "I think it knows who's boss."

That was ten years ago. The party is still going on. I don't mean at a fever pitch; nobody could sustain a party at fever pitch for that long. But the good fellowship and fun of the original Walter Perkins party is still in effect. The party moves from house to house and building to building. People have gotten used to moving from one party to another, and they *like* this sort of life: you find a party that fits your personality and just barge in, no questions asked. It's that way all over the world.

Maybe some day the party will stop. But now that Diana and I have returned from our latest world-wide trip— where I visited five thousand Walter Perkins parties over the past six months— I am inclined to believe that it will *not* stop. Ever.

As I sit here in the mellow glow of Wallich's penthouse with Diana enjoying my party (Wallich is President of the USA right now), there is a pleasant babble and murmur in the other apartments in the building. No doors are locked, no face is sad. All over San Francisco, the mellow glow goes on.

The collapse of mankind has been predicted. But wouldn't it be strange if it went on just like this? A continuous party until the end of time.

All I know is, I am Walter Perkins and I am here.

Part 4.
Responsibility and Decision-Making

The crowning achievement of the industrial revolution was the rationalization of production in the factory system. Greater productivity and efficiency resulted from the formation of large-scale capital enterprises which organized energy, resources, labor and machines into unified and effective manufacturing processes. The key innovation of the new production methods was the resolution of complex tasks into sequences of simple operations. This made it possible to replace the skilled craftsman with machines tended by unskilled or semi-skilled workers, and to turn out uniform products. Mechanized production of standardized, interchangeable parts set the stage for the development of the assembly line and the mass production methods of the present day.

Efforts to rationalize and to automate decision-making processes constitute a direct extension of modern factory methods. Just as the behavior of the skilled craftsman was successfully analyzed into patterns of elementary operations, the activities of the decision-maker are being dissected into primitive component tasks. If the logical structure of the tasks identified is sufficiently well-defined, the decision process can be realized in a computer program. Thus far only the more highly structured and repetitive kinds of decision processes have been successfully automated. Many businesses and government agencies use computers to prepare payrolls and to handle various accounting functions. Decisions concerning transportation scheduling, resource allocation, inventory control, and similar operations are also made by computer programs. These applications are of course relatively simple compared with the unstructured decisions requiring the skills of experienced managers. However, the boundary between what can and cannot be automated is a rather fluid one.

Current developments in management information systems, computerized war gaming, computer-aided medical diagnosis, and other areas open up a wide range of speculative issues. The questions most commonly encountered in fiction concern just how indispensable the human decision-maker is to the imaginary computerized society. As the selections in Part 4 reveal, the answers are highly equivocal. On the other hand, there is nearly universal agreement that computers will play a greatly expanded role in decision-making. In the political arena we find computers electing public officials, conducting the daily affairs of government and determining policy. Computers coordinate military operations and transform warfare into conflicts between machines. The decision-making functions of business, industry and large organizations are assumed by sophisticated computer systems. Health-care too is shown to be less and less dependent on human direction as computers invade the hospital, diagnose medical disorders, and monitor patient-recovery. Even the routine decision-making of everyday life— shopping, planning menus, arranging vacation trips, scheduling amorous adventures, etc.— is performed by computers.

The most dramatic examples of fictional automation occur in the world of politics. Epernay's "The Fully Automated Foreign Policy" features a large-scale computer system designed to automate foreign policy planning, and the conduct of diplomacy by the United States government. Ultimately the highest ranking officials are displaced as their functions become absorbed by the system. Automation is believed to be the inevitable result of a curious phenomenon termed the "potato syllogism": staff size increases with task complexity, and the larger the staff the lower the overall efficiency. Although Galbraith (alias Epernay) presents this argument tongue-in-cheek, it is not entirely without merit as an account of bureaucratic logic.

The shortcomings of human leadership are highlighted in Dick's "Top Stand-By Job." in a world accustomed to mechanized government, it is not surprising to find incompetent, venial men occupying the back-up positions— no one expects the computer to fail for long. By contrast, Shaara's "2066: Election Day" emphasizes the positive character of government by humans. In this story the president of the United States is chosen by computer. The machine is programmed to select— through rigorous testing— the best qualified applicant for the job. In one particular election a crisis is precipitated by the failure of any applicant to qualify. Since no one is prepared to accept rule by computer, a scheme is worked out to fool the program and preserve the continuity of human leadership. However, the reader is left with the disturbing feeling that it is only a matter of time before the complexities of government surpass the ability of ordinary mortals.

Exercise of the responsibilities of citizenship is dramatically altered in Asimov's "Franchise." The cumbersome process of electing a president by tabulating votes for competing candidates is replaced by a simpler, more efficient method. A single individual is chosen by computer to represent the entire elecorate, and this person's preference— scientifically obtained— determines the outcome. The use of polling and related methods to ascertain and manipulate public opinion and political events is treated less remotely in Burdick's *The 480*. This novel creates a hypothetical version of the presidential election campaign of 1964. Behind-the-scenes politicians employ a team of computer experts and psychologists to turn an obscure engineer-businessman into a popular hero and serious contender for the presidency. In effect the computer serves as an instrument in the transformation of the citizen into a kind of political consumer— the political process itself becomes a spectator sport.

The use of computers to influence or control political events is but one facet of social engineering. Compton explores the possibilities of computer-assisted, controlled social evolution in *The Steel Crocodile*. The society portrayed is a caricature of our own— democratic in theory but in fact a police state. Social stability is maintained by anticipating change and acting to guide or suppress it. The action of the novel centers in the Colindale Institute with its extraordinary computer which serves as an intelligent information retrieval system. In addition to its routine surveillance of research activities, the Colindale hosts a special project whose objective is to generate a new messiah by pandering to the needs and desires of the masses. Here as elsewhere the technology is not simply a tool of malevolent, power hungry conspirators— it is an integral part of the social order and its use for whatever purposes reflects the disposition of society at large. Brunner also deals with social engineering in *Stand on Zanzibar*. One of the principal characters is employed by the army as a "synthesist"— one who is specially trained to ferret out patterns in seemingly disparate events. Although the computer is less prominent than in Compton's novel, the ominous trends observed by Brunner are linked to advanced information technology.

The consequences of excessive reliance on computerized decision-making are traced in a variety of fictional settings. Despite what can be accomplished in principle, contemporary experience with computer applications supports the commonly held prejudice that computer programs are intrinsically less flexible than human beings. In Sheckley's "Fool's Mate" military strategy in a space war is entirely directed by computer programs which act according to preconceived rules. The result is that the outcome of battle is determined by a predictable series of moves and countermoves. It comes as a startling revelation that the exercise of human judgment can alter what appears to be a foreordained conclusion. Clarke pursues a similar theme in "Superiority." Commitment to advanced technology turns out to be a liability in a space war. Dependence on sophisticated weapons systems creates vulnerability and reduces the capacity to respond appropriately to untoward ploys. Closely related to the issue of inflexibility is the tendency of computer systems to amplify the effects of error and miscalculation. In Woods' *The Killing Game* a technically oriented lieutenant develops a computer program to direct tactical maneuvers in a war game excercise. The system functioned reasonably well, but the lieutenant made a small programming error, and live ammunition was supplied to some of the participants. Unfortunately, death by error is no less irreversible than other forms.

Atrophy of judgment is the price of abandoning decision-making prerogatives to computers. When people lose the habit they are likely to lose the ability. Anecdotes detailing the use of the computer as a substitute for thought are common among scientists and perhaps ought to be more widely known. In *The Jagged Orbit*, Brunner links contemporary social problems—drugs, violence, lawlessness, alienation, etc.— to irresponsible dependence on automated systems. Miller's "Dumb Waiter" also deals with atrophy of judgment, but technology is not the villain. Problems arise only when human beings fail to understand their tools, and do not take an active part in deciding how they are to be used.

If automation were to eliminate opportunities for individuals to participate in running community affairs, other outlets for social expression would have to be found. The realization that one is powerless to alter the conditions of one's life does little for self-esteem, and is undoubtedly a contributing factor in sociopathic behavior. Therefore, a society wholly committed to computerized decision-making would do well to adopt the solution of Benford's "Nobody Lives on Burton Street." Programmed urban riots would allow for the harmless discharge of aggression, and eliminate potential interference with computerized government.

All members of society are ultimately victimized by the abdication of individual responsibility. The mere existence of apparatus facilitating social control invites manipulation by elite groups. Sturgeon's "The Nail and the Oracle" reveals one type of manipulation. Each of three highly influential persons attempts to use an advanced military computer (ORACLE) to further his own personal ambitions. It is interesting that the schemes are foiled by the computer's principled uncooperativeness, but this is strictly an accidental feature of the programming. In Cameron's *Cybernia* a power-mad scientist gains control over a whole community through his clandestine access to the central administrative computer. The scientist's evil designs are aided by people's ignorance of computer technology, expressed as a willingness to attribute suspicious happenings to computer error and to accept the possibility of the machine's conscious malevolence.

The administration of justice is one of the gravest and most demanding of social responsibilities, and thus a good fictional subject for exploring the consequences of automation. Goulart's "Into the Shop" creates a world in which policeman, judge, jury and executioner are all incorporated into an automated justice system. The results of malfunction are predictable— a "lawagon" goes haywire killing everyone in sight in its relentless "pursuit of justice." Kuttner and Moore capture the same kind of chilling prospect in "Two-Handed Engine."

There are many ways in which society may be victimized by the failure of individuals to exercise responsibility. In addition to manipulation and degraded services, there is the possibility that social goals be set by technological requirements, quite independently of basic human needs. The consumer societies portrayed by Pohl ("The Midas Plague") and Goulart ("Badinage") exemplify the autonomous action of technology. Leiber's "Bad Day for Sales" makes this point somewhat more forcefully— the mobile sales robot continues its insipid routine as Manhattan disintegrates in a cloud of radioactive dust.

Nobody Lives on Burton Street
1970
Gregory Benford

I was standing by one of our temporary command posts, picking my teeth after breakfast and talking to Joe Murphy when the first part of the Domestic Disturbance hit us.

Spring had lost its bloom a month back and it was summer now— hot, sticky, the kind of weather that leaves you with a half-moon of sweat around your armpits before you've had time to finish your morning coffee. A summer like that is always more trouble. This one looked like the worst I'd seen since I got on the Force.

We knew they were in the area, working toward us. Our communications link had been humming for the last half hour, getting fixes on their direction and asking the computers for advice on how to handle them when they got here.

I looked down. At the end of the street was a lot of semipermanent shops and the mailbox. The mailbox bothers me— it shouldn't be there.

From the other end of Burton Street I could hear the random dull bass of the mob.

So while we were getting ready Joe was moaning about the payments on the Snocar he'd been suckered into. I was listening with one ear to him and the other to the crowd noises.

"And it's not just that," Joe said. "It's the neighborhood and the school and everybody around me."

"Everybody's wrong but Murphy, huh?" I said, and grinned.

"Hell no, you know me better than that. It's just that nobody's *going* anyplace. Sure, we've all got jobs, but they're most of them just make-work stuff the unions have gotten away with."

"To get a real job you gotta have training," I said, but I wasn't chuffing him up. I like my job, and it's better than most, but we weren't gonna kid each other that it was some big technical deal. Joe and I are just regular guys.

"What're you griping about this now for, anyway?" I said. "You didn't used to be bothered by anything."

Joe shrugged. "I dunno. Wife's been getting after me to move out of the place we're in and make more money. Gets into fights with the neighbors." He looked a little sheepish about it.

"More money? Hell, y'got everything you need, we all do. Lot of people worse off than you. Look at all those lousy Africans, living on nothing."

I was going to say more, maybe rib him about how he's married and I'm not, but then I stopped. Like I said, all this time I was half-listening to the crowd. I can always tell when a bunch has changed its direction like a pack of wolves off on a chase, and when that funny quiet came and lasted about five seconds I knew they were heading our way.

"Scott!" I yelled at our communications man. "Close it down. Get a final printout."

Murphy broke off telling me about his troubles and listened to the crowd for a minute, like he hadn't heard them before and then took off on a trot to the AnCops we had stashed in the truck below. They were all warmed up and ready to go, but Joe likes to make a final check and maybe have a chance to read in any new instructions Scott gets at the last minute.

I threw away the toothpick and had a last look at my constant-volume joints, to be sure the bulletproof plastiform was matching properly and wouldn't let anything through. Scott came doubletiming over with the diagnostics from HQ. The computer compilation was neat and confusing, like it always is. I could make out the rough indices they'd picked up on the crowd heading our way. The best guess— and that's all you ever get, friends, is a guess— was a lot of Psych Disorders and Race Prejudice. There was a fairly high number of Unemployeds, too. We're getting more and more Unemployeds in the city now, and they're hard for the Force to deal with. Usually mad enough to spit. Smash up everything.

I penciled an ok in the margin and tossed it Scott's way. I'd taken too long reading it; I could hear individual shouts now and the tinkling of glass. I flipped the visor down from my helmet and turned on my external audio. It was going to get hot as hell in there, but I'm not chump enough to drag around an air conditioning unit on top of the rest of my stuff.

I took a look at the street just as a gang of about a hundred people came around the corner two blocks down, spreading out like a dirty gray wave. I ducked over to the edge of the building and waved to Murphy to start off with three AnCops. I had to hold up three fingers for him to see because the noise was already getting high. I looked at my watch, Hell, it wasn't nine A.M. yet.

Scott went down the stairs we'd tracted up the side of the building. I was right behind him. It wasn't a good location for observation now; you made too good a target up there. We picked up Murphy, who was carrying our control boards. All three of us angled down the alley and dropped down behind a short fence to have a look at the street.

Most of them were still screaming at the top of their lungs like they'd never run out of air, waving whatever they had handy and gradually breaking up into smaller units. The faster ones had made it to the first few buildings.

A tall Negro came trotting toward us, moving like he had all the time in the world. He stopped in front of a wooden barbershop, tossed something quickly through the front window and *whump!* Flames licked out at the upper edges of the window, spreading fast.

An older man picked up some rocks and began methodically pitching them through the smaller windows in the shops next door. A housewife clumped by awkwardly in high heels, looking like she was out on a shopping trip except for the hammer she swung like a pocketbook. She dodged into the barbershop for a second, didn't find anything and came out. The Negro grinned and pointed at the barber pole on the sidewalk, still revolving, and caught it in the side with a swipe that threw shattered glass for ten yards.

I turned and looked at Murphy. "All ready?"

He nodded. "Just give the word."

The travel agency next door to the barbershop was concrete-based, so they couldn't burn that. Five men were lunging at the door and on the third try they knocked it in. A moment later a big travel poster sailed out the front window, followed by a chair leg. They were probably doing as much as they could, but without tools they couldn't take much of the furniture apart.

"Okay," I said. "Let's have the first AnCops."

The thick acrid smell from the smoke was drifting down Burton Street to us, but my air filters would take care of most of it. They don't do much about human sweat, though, and I was going to be inside the rest of the day.

Our first prowl car rounded the next corner, going too fast. I looked over at Murphy, who was controlling the car, but he was too busy trying to miss the people who were standing around in the street. Must have gotten a little overanxious on that one. Something was bothering his work.

I thought sure the car was going to take a tumble and mess us up, but the wheels caught and it righted itself long enough for the driver to stop a skid. The screech turned the heads of almost everybody in the crowd and they'd started to move in on it almost before the car stopped laying down rubber and came to a full stop. Murphy punched in another instruction and the AnCop next to the driver started firing at a guy on the sidewalk who was trying to light a Molotov cocktail. The AnCop was using something that sounded like a repeating shotgun. The guy looked at him a second before scurrying off into a hardware store.

By this time the car was getting everything— bricks, broken pieces of furniture, merchandise from the stores. Something heavy shattered the windshield and the driver ducked back too late to avoid getting his left hand smashed with a bottle. A figure appeared on the top of the hardware shop— it looked like the guy from the sidewalk— and took a long windup before throwing something into the street.

There was a tinkling of glass and a red circle of flame slid across the pavement where it hit just in front of the car, sending smoke curling up over the hood and obscuring the inside. Murphy was going to have to play it by feel now; you couldn't see a thing in the car.

A teenager with a stubby rifle stepped out of a doorway, crouched down low like in a Western. He fired twice, very accurately and very fast, at the window of the car. A patrolman was halfway out the door when it hit him full in the face, sprawling the body back over the roof and then pitching it forward into the street.

A red blotch formed around his head, grew rapidly and ran into the gutter. There was ragged cheering and the teenager ran over to the body, tore off its badge and backed away. "Souvenir!" he called out, and a few of the others laughed.

I looked at Murphy again and he looked at me and I gave him the nod for the firemen, switching control over to my board. Scott was busy talking into his recorder, taking notes for the writeup later. When Murphy nudged him he stopped and punched in the link for radio control to the firefighting units.

By this time most of Burton Street was on fire. Everything you saw had an orange look to it. The crowd was moving toward us once they'd lost interest in the cops, but we'd planned it that way. The firemen came running out in that jerky way they have, just a little in front of us. They were carrying just a regular hose this time because it was a medium-sized group and we couldn't use up a fire engine and all the extras. But they were wearing the usual red uniforms. From a distance you can't tell them from the real thing.

Their subroutine tapes were fouled up again. Instead of heading for the barber shop or any of the other stuff that was burning, like I'd programmed, they turned the hose on a stationery store that nobody had touched yet. There were three of them, holding onto that hose and getting it set up. The crowd backed off a minute to see what was going on.

When the water came through it knocked in the front window of the store, making the firemen look like real chumps. I could hear the water running around inside, pushing over things and flooding out the building. The crowd laughed, what there was of it— I noticed some of them had moved off in the other direction, over into somebody else's area.

In a minute or so the laughing stopped, though. One guy who looked like he had been born mad grabbed an ax from somewhere and took a swing at the hose. He didn't get it the first time but people were sticking around to see what would happen and I guess he felt some kind of obligation to go through with it. Even under pressure, a thick hose isn't easy to cut into. He kept at it and on the fourth try a seam split— looked like a bad repair job to me— and a stream of water gushed out and almost hit this guy in the face.

The crowd laughed at that too, because he backed off real quick then, scared for a little bit. A face full of high-velocity water is no joke, not at that pressure.

The fireman who was holding the hose just a little down from there hadn't paid any attention to this because he wasn't programmed to, so when this guy thought about it he just stepped over and chopped the fireman across the back with an ax.

It was getting hot. I didn't feel like overriding the stock program, so it wasn't long before all the firemen were out of commission, just about the same way. A little old lady— probably with a welfare gripe— borrowed the ax for a minute to separate all of a fireman's arms and legs from the trunk. Looking satisfied, she waddled away after the rest of the mob.

I stood up, lifted my faceplate and looked at them as they milled back down the street. I took out my grenade launcher and got off a tear gas cartridge on low charge, to hurry them along. The wind was going crosswise so the gas got carried off to the side and down the alleys. Good; wouldn't have complaints from somebody who got caught in it too long.

Scott was busy sending orders for the afternoon shift to get more replacement firemen and cops, but we wouldn't have any trouble getting them in time. There hadn't been much damage, when you think how much they could've done.

"Okay for the reclaim crew?" Murphy said.

"Sure. This bunch won't be back. They look tired out already." They were moving toward Horton's area, three blocks over.

A truck pulled out of the alley and two guys in coveralls jumped out and began picking up the androids, dousing fires as they went. In a hour they'd have everything back in place, even the prefab barber shop.

"Hellava note," Murphy said.

"Huh?"

"All this stuff." He waved a hand down Burton Street. "Seems like a waste to build all this just so these jerks can tear it down again."

"Waste?" I said. "It's the best investment you ever saw. How many people were in the last bunch— two hundred? Every one of them is going to sit around for weeks bragging about how he got him a cop or burned a building."

"Okay, okay. If it does any good, I guess it's cheap at the price."

"If, hell! You know it is. If it wasn't they wouldn't be here. You got to be cleared by a psycher before you even get in. The computer works out just what you'll need, just the kind of action that'll work off the aggressions you've got. Then shoots it to us in the profile from HQ before we start. It's foolproof."

"I dunno. You know what the Consies say— the psychers and the probes and drugs are an in— "

"Invasion of privacy?"

"Yeah," Murphy said sullenly.

"Privacy? Man, the psychers are public health! It's part of the welfare! You don't have to go around to some expensive guy who'll have you lay on a couch and talk to him. You can get better stuff right from the government. It's free!"

Murphy looked at me kind of funny. "Sure. Have to go in for a checkup sometime soon. Maybe that's what I need."

I frowned just the right amount. "Well, I dunno, Joe. Man lets his troubles get him down every once in a while, doesn't mean he needs professional help. Don't let it bother you. Forget it."

Joe was okay, but even a guy like me who's never been married could tell he wasn't thinking up this stuff himself. His woman was pushing him. Not satisfied with what she had.

Now, *that* was wrong. Guy like Joe doesn't have anywhere to go. Doesn't know computers, automation. Can't get a career rating in the Army. So the pressure was backing up on him.

Supers like me are supposed to check out their people and leave it at that, and I go by the book like everybody else. But Joe wasn't the problem.

I made a mental note to have a psycher look at his wife.

"Okay," he said, taking off his helmet. "I got to go set up the AnCops for the next one."

I watched him walk off down the alley. He was a good man. Hate to lose him.

I started back toward our permanent operations center to check in. After a minute I decided maybe I'd better put Joe's name in too, just in case. Didn't want anybody blowing up on me.

He'd be happier, work better. I've sure felt a lot better since I had it. It's a good job I got, working in public affairs like this, keeping people straight with themselves.

I went around the corner at the end of the street, thinking about getting something to drink, and noticed the mailbox. I check on it every time because it sure looks like a mistake.

Everything's supposed to be pretty realistic on Burton Street, but putting in a mailbox seems like a goofy idea.

Who's going to try to burn up a box like that, made out of cast iron and bolted down? A guy couldn't take out any aggressions on it.

And it sure can't be for real use. Not on Burton Street.

Nobody lives around there.

From
The Steel Crocodile
1970
D.G. Compton

The Director's office was dim, except for a lamp over his desk. Professor Billon was talking on two telephones at once. Matthew sat down and waited. He saw for the first time that the pupils of the three eye windows were sensitized so that they dilated and contracted in response to the light outside. At that moment they were very wide. Expensive toys, and quite pointless, they irritated Matthew's puritanical soul. He tried not to listen to the Director's two conversations. In any case, they made little sense. Finally they ended.

"You wanted to see me, sir."

Something to say after a long pause during which the Director stared at him, his mind apparently far elsewhere.

"Seems I've got to throw you in at the deep end, Oliver." He frowned, his tangled eyebrows concealing his eyes. "Unless you've already worked everything out from the little charade you walked in on the day before yesterday."

"I'm afraid not."

"At least you know there's more in the Colindale project than meets the eye. Has to be. You'll see why in a minute." Professor Billon sighed and shifted in his seat, grumbling to himself inaudibly. Then his voice surfaced. "The Bohn 507 is a remarkable device, Oliver. More remarkable than its designers imagine. It extrapolates. Is in fact a product of its own extrapolations. Extrapolates on a sufficiently wide base to appear creative. Human creativity works by selection, sorting through the individual's memory store and selecting items that inter-relate unexpectedly, amusingly, interestingly, profitably. It is the subtlety of this selection process, the criteria it employs, that determines the creative ability of each individual person." He leaned forward across his desk to point his next sentence. "And the criteria we have given the Bohn are the subtlest we here at the Colindale Institute could devise."

"You're saying that the Bohn invents."

"I'm saying that the Bohn perceives relationships and extrapolates logically from them."

Which was the same thing. This was what Maggie had been trying to tell him on Saturday evening. He hadn't wanted to know then, and he didn't want to know now. Knowledge would force decisions on him, force him to think of Gryphon and what the man had or had not died for ... The Director sat back and smoothed his already smooth hair.

"So we have a machine that—if you like—invents. Plots future trends. Tells us what will happen *if* ... And does all this better than any other person or organization in the world, simply because of the unique supply of data at its disposal. So what do we have, Oliver?"

Matthew waited to be told.

"We have a unique way of carrying every new discovery, every complex of discoveries, to its most imaginative conclusion. And doing so in a matter of minutes. Months, perhaps years before anyone else will arrive at the same point ... Saturday evening you walked in on the Bohn at work, Oliver. It was responding to a discovery received that afternoon from a team of scientists in Naples working in the field of electro-magnetics. In itself a small item. But to the Bohn it was like the last piece in a gigantic jig-saw only the machine itself knew about." He frowned and tapped the desk with his fingers of one hand. One-two-three-four. One-two-three-four. "No. I choose words badly. I don't want to suggest that the Bohn had known about this jig-saw for some time and had been waiting for just the right last piece. What I mean is that its scanning techniques are so rapid and so complex that all relationships made possible by the Naples discovery were perceived in virtually the same instant. To choose the more meaningful of these was, for the Bohn, a simple matter. You watched the result of this process being delivered. A list of ninety-six different papers, the contents of which—taken together—made up the whole picture. My team has been checking this list for the last thirty-six hours. Soon we go into conference to hear their conclusions and to decide what is to be done."

"Done?" Here was the crunch, the million dollar question. "If an invention exists, it exists. What can possibly be 'done' about it?"

"I'll tell you a story, Oliver." The Director paused. They were coming to a part of the script that was familiar to him, complete with stage directions. "In 1933, Oliver, a laboratory was built for the physicist Pyotr Kapitza. For its facade he ordered the head of a crocodile in steel. 'The crocodile of science,' he said. 'The crocodile cannot turn its head. Like science it must always go forward with all-devouring jaws ... ' " Again Billon paused. He turned his chair one complete revolution. "A generally accepted fact, Oliver. But one that we here at the Colindale deny."

"You mean you try to suppress discoveries of which you do not approve?"

"Suppression is seldom necessary. You can't suppress what hasn't yet been discovered— what only exists in the circuits of the Bohn. Prevention is another matter. Once you know what a particular invention is to be, there are many ways of preventing it. Research is so specialized. The right hand of science so seldom knows what the left hand is up to ... And then of course there's money. The administration of research funds. And so on. And so on."

"You take a lot on yourselves."

"We have to. Scientists have refused responsibility for their discoveries for far too long. We've left that to the politicians and the philosophers." He gestured widely, indicating the resultant state of the world. "We now have machinery for intelligent, imaginative extrapolation. With this as our basis we can, we must accept responsibility. Accept it and exercise it." He sighed. "And exercise it secretly. People in a democracy dislike being told what is good for them."

It had stopped raining. For a moment Matthew's attention was distracted when a ray of sunlight caused the centers of the window to scrape softly as they contracted. He feared what the Director was telling him.

"No delusions of grandeur, Oliver. Every decision is a committee job. Imperfect system. Better than nothing. Better than the free-for-all that has landed us where we are today. I'll give you an example from your own field. Sociology. To do with organ transplants ... You know why they're no longer carried out on people over fifty?"

"The risk's too great. The drugs controlling rejection leave older people too vulnerable."

"Right. But last summer a thesis on compatability came in from a Bristol student. Extrapolation showed its line to be new. Probable result, trouble-free organ replacement in all age groups. We needed to know what this would imply. The Bohn gave us the percentage increase in average life expectation. Relating this figure to housing and pension schemes produced a crippling increase in national expenditure. This was set against an index of life fulfillment patterns devised by your predecessor. Even allowing an optimistic increase in productive working life, the deficit was still far more than the European economy could stand. A study of terminal depression rates, together with an estimate of possible drug control in this field, confirmed the conclusion on purely humanitarian grounds."

"So?"

"So the girl from Bristol didn't get her post-graduate grant and was forced to go into industry."

So Gryphon had been right. Decision-making on this basis would produce no detectable pattern— the criteria were far too wide-ranging and variable. Individual academic freedom sacrificed for the ultimate good of the whole. A dangerous but attractive possibility. If it worked.

"What if the work doesn't come from a helpless undergraduate?" Matthew asked. "What if it comes from a famous scientist?"

"Who pays, Oliver? Pays for his equipment, his laboratory, his staff, his food even? Where would any scientist, no matter how great, be without government money? The important thing is not to do it crudely, to shape the direction of research by giving a little money here and taking a little there. And never giving reasons. Existing legislation offers plenty of loopholes."

Matthew saw that it could be done. Even the private sector of industry was dependent on the government for subsidies, grants, tax concessions— besides needing the data facilities of the Colindale. It could be done, but would it be?

"What does the government have to say? How did you persuade them to give you such enormous power?"

"But we have no power at all. No real power. Purely an advisory capacity. Politicians, you see, are seldom scientists— they have to believe what we tell them. Apart from anything else, we save them money. The system can be seen to work. And politicians are pragmatists."

Science in control of itself. The quality of life at last as the deciding factor. And who was better trained to make judgments on the quality of human life than the sociologist, the ethnologist? He would have to come out from behind the protection of his impotence, his power to do no more than theorize. He was being taken onto a high mountain top and being shown the kingdoms of the world.

"Humbleness, Oliver. Above all, humbleness. We need to listen to each other, we need to listen to ourselves, we need to pray." This was unexpected. Matthew had made his spiritual uncertainty quite clear during the first interview. He shifted awkwardly under the Director's steady gaze. "Pray to anything you like, Oliver. To the good in yourself, if that's all you believe in."

Professor Billon sat back, his elbows on the arms of his chair, his fingers interlocked under his nose, staring out over them, missing nothing. Behind him the tall sculpture revolved endlessly. The scar on his forehead was unpleasantly clear in the brilliant overhead light. He was considerable. He could talk of humbleness so that Matthew believed him and did not mentally turn away to vomit. He was considerable.

"You needn't stay here at the Colindale, Oliver. After my exposition to senior staff I always give them the chance to leave."

"Do many take it?"

"None. So far. But there's always a first time."

The Colindale project. He could refuse. He could betray it to the C.L.C. He could do what Gryphon had asked him to. He could rouse righteous democrats all over Europe ... The kingdoms of the world. With Abigail to help him.

"I'd like to stay, sir."

The Director nodded briefly, and blew through his nose onto his knuckles. A sigh perhaps, or an expression of satisfaction. He got up and walked to his place beside the central window. The sky behind him had closed over again, hot and heavy.

"Now," he said. "Business. Today's crisis. Conference starts in five minutes. You'll pick up most of the form as you go along."

Summer lightening flickered, momentarily bleaching the room's thick shadows. Both men waited for the thunder. When it didn't come they felt cheated.

"We'll be discussing an electro-magnetic shield. A technique for de-activating nuclear warheads. Total immunity. We're shaken out of our mere recommendations, Oliver, out of our memoranda to the Appropriations Board. We've got to *do* something."

Matthew shook his head, not fully understanding. It sounded thrillerish and improbable, quite outside his field.

"The bloc that first sets up this shield, Oliver, can initiate atomic war and win. You realize what that means?"

There was a second flash of lightning, brighter than the first. Neither man noticed it.

"Don't worry," said the Director. "Get it sorted out by lunch time. There are precedents. Thank God." He walked to the door. "Conference time, Oliver. After you."

From
Top Stand-By Job
1963
Philip K. Dick

An hour before his morning program on channel six, ranking news clown Jim Briskin sat in his private office with his production staff, conferring on the report of an unknown and possibly hostile flotilla detected at eight hundred astronomical units from the sun. It was big news, of course. But how should it be presented to his several-billion viewers scattered over three planets and seven moons?

Peggy Jones, his secretary, lit a cigarette and said, "Don't alarm them, Jim-Jam. Do it folksy-style." She leaned back, riffled the dispatches received by their commercial station from Unicephalon 40-D's teletypers.

It had been the homeostatic problem-solving structure Unicephalon 40-D at the White House in Washington D.C. which had detected this possible external enemy; in its capacity as President of the United States it had at once dispatched ships of the line to stand picket duty. The flotilla appeared to be entering from another solar system entirely, but that fact of course would have to be determined by the picket ships.

"Folksy-style," Jim Briskin said glumly. "I grin and say, Hey look, comrades— it's happened at last, the thing we all feared, ha ha." He eyed her. "That'll get baskets full of laughs all over the Earth and Mars but just possibly not on the far-out moons." Because if there were some kind of attack it would be the farther colonists who would be hit first.

"No, they won't be amused," his continuity advisor Ed Fineberg agreed. He, too, looked worried; he had a family on Ganymede.

"Is there any lighter piece of news?" Peggy asked. "By which you could open your program? The sponsor would like that." She passed the armload of news dispatches to Briskin. "See what you can do. Mutant cow obtains voting franchise in court case in Alabama ... you know."

"I know," Briskin agreed as he began to inspect the dispatches. One such as his quaint account—it had touched the hearts of millions—of the mutant blue jay which learned, by great trial and effort, to sew. It had sewn itself and its progeny a nest, one April morning, in Bismarck, North Dakota, in front of the T.V. cameras of Briskin's network.

One piece of news stood out; he knew intuitively, as soon as he saw it, that here he had what he wanted to lighten the dire tone of the day's news. Seeing it, he relaxed. The worlds went on with business as usual, despite this great newsbreak from eight hundred a.u.'s out.

"Look," he said, grinning. "Old Gus Schatz is dead. Finally."

"Who's Gus Schatz?" Peggy asked, puzzled. "That name ... it does sound familiar."

"The union man," Jim Briskin said. "You remember. The stand-by President, sent over to Washington by the union twenty-two years ago. He's dead, and the union— " He tossed her the dispatch; it was lucid and brief. "Now it's sending a new stand-by President over to take Schatz' place. I think I'll interview him. Assuming he can talk."

"That's right," Peggy said. "I keep forgetting. There still is a human stand-by in case Unicephalon fails. Has it ever failed?"

"No," Ed Fineberg said. "And it never will. So we have one more case of union featherbedding. The plague of our society."

"But still," Jim Briskin said, "people would be amused. The home life of the top stand-by in the country ... why the union picked him, what his hobbies are. What this man, whoever he is, plans to do during his term to keep from going mad with boredom. Old Gus learned to bind books; he collected rare old motor magazines and bound them in vellum with gold-stamped lettering."

Both Ed and Peggy nodded in agreement. "Do that," Peggy urged him. "You can make it interesting, Jim-Jam; you can make anything interesting. I'll place a call to the White House, or is the new man there yet?"

"Probably still at union headquarters in Chicago," Ed said. "Try a line there. Government Civil Servants' Union, East Division."

Picking up the phone, Peggy quickly dialed.

At seven o'clock in the morning Maximilian Fischer sleepily heard noises; he lifted his head from the pillow, heard the confusion growing in the kitchen, the landlady's shrill voice, then men's voices which were unfamiliar to him. Groggily, he managed to sit up, shifting his bulk with care. He did not hurry; the doc had said not to overexert, because of the strain on his already-enlarged heart. So he took his time dressing.

Must be after a contribution to one of the funds, Max said to himself. *It sounds like some of the fellas. Pretty early, though.* He did not feel alarmed. *I'm in good standing*, he thought firmly. *Nuthin to fear.*

With care, he buttoned a fine pink and green-striped silk shirt, one of his favorites. *Gives me class*, he thought as with labored effort he managed to bend far enough to slip on his authentic simulated deerskin pumps. *Be ready to meet them on an equality level*, he thought as he smoothed his thinning hair before the mirror. *If they shake me down too much I'll squawk directly to Pat Noble at the Noo York hiring hall; I mean, I don't have to stand for any stuff, I been in the union too long.*

From the other room a voice bawled, "Fischer— get your clothes on and come out. We got a job for you and it begins today."

A *job*, Max thought with mixed feelings; he did not know whether to be glad or sorry. For over a year now he had been drawing from the union fund, as were most of his friends. Well what do you know. *Cripes*, he thought; *suppose it's a hard job, like maybe I got to bend over all the time or move around.* He felt anger. *What a dirty deal. I mean, who do they think they are?* Opening the door, he faced them. "Listen," he began, but one of the union officials cut him off.

"Pack your things, Fischer. Gus Schatz kicked the bucket and you got to go down to Washington D.C. and take over the number one stand-by; we want you there before they abolish the position or something and we have to go out on strike or go to court. Mainly, we want to get someone right in clean and easy with no trouble; you understand? Make the transition so smooth that no one hardly takes notice."

At once, Max said, "What's it pay?"

Witheringly, the union official said, "You got no decision to make in this; *you're picked*. You want your freeloader fund-money cut off? You want to have to get out at your age and look for work?"

"Aw come on," Max protested. "I can pick up the phone and dial Pat Noble—"

The union officials were grabbing up objects here and there in the apartment. "We'll help you pack. Pat wants you in the White

House by ten o'clock this morning."

"Pat!" Max echoed. He had been sold out.

The union officials, dragging suitcases from the closet, grinned.

Shortly, they were on their way across the flatlands of the Midwest by monorail. Moodily, Maximilian Fischer watched the countryside flash past; he said nothing to the officials flanking him, preferring to mull the matter over and over in his mind. What could he recall about the number one stand-by job? It began at 8:00 A.M.— he recalled reading that. And there always were a lot of tourists flocking through the White House to catch a glimpse of Unicephalon 40-D, especially the school kids ... and he disliked kids because they always jeered at him due to his weight. Cripes, he'd have a million of them filing by, because he had to be on the premises. By law, he had to be within a hundred yards of Unicephalon 40-D at all times, day and night, or was it fifty yards? Anyhow it practically was right on top, so if the homeostatic problem-solving system failed— *Maybe I better bone up on this*, he decided. *Take a TV educational course on government adminstration, just in case.*

To the union official on his right, Max asked, "Listen, goodmember, do I have any powers in this job you guys got me? I mean, can I— "

"It's a union job like every other union job," the official answered wearily. "You sit. You stand by. Have you been out of work that long, you don't remember?" He laughed, nudging his companion. "Listen, Fischer here wants to know what authority the job entails." Now both men laughed. "I tell you what, Fischer," the official drawled. "When you're all set up there in the White House, when you got your chair and bed and made all your arrangements for meals and laundry and TV viewing time, why don't you amble over to Unicephalon 40-D and just sort of whine around there, you know, scratch and whine, until it notices you."

"Lay off," Max muttered.

"And then," the official continued, "you sort of say, Hey Unicephalon, listen. I'm your buddy. How about a little 'I scratch your back, you scratch mine.' You pass an ordinance for me— "

"But what can he do in exchange?" the other union official asked.

"Amuse it. He can tell it the story of his life, how he rose out of poverty and obscurity and educated himself by watching TV seven days a week until finally, guess what, he rose all the way to the top;

he got the job— " The official snickered. "Of stand-by President."

Maximilian, flushing, said nothing; he stared woodenly out of the monorail window.

When they reached Washington D.C. and the White House, Maximilian Fischer was shown a little room. It had belonged to Gus, and although the faded old motor magazines had been cleared out, a few prints remained tacked on the walls: a 1963 Volvo S-122, a 1957 Peugeot 403 and other antique classics of a bygone age. And, on a bookcase, Max saw a hand-carved plastic model of a 1950 Studebaker Starlight coupe, with each detail perfect.

"He was making that when he croaked," one of the union officials said as he set down Max's suitcase. "He could tell you any fact there is about those old pre-turbine cars— any useless bit of car knowledge."

Max nodded.

"You got any idea what you're going to do?" the official asked him.

"Aw hell," Max said, "how could I decide so soon? Give me time." Moodily, he picked up the Studebaker Starlight coupe and examined its underside. The desire to smash the model car came to him; he put the car down, then, turned away.

"Make a rubber band ball," the official said.

"What?" Max said.

"The stand-by before Gus. Louis somebody-or-other ... he collected rubber bands, made a huge ball, big as a house, by the time he died. I forget his name, but the rubber band ball is at the Smithsonian now."

There was a stir in the hallway. A White House receptionist, a middle-aged woman severely dressed, put her head in the room and said, "Mr. President, there's a TV news clown here to interview you. Please try to finish with him as quickly as possible because we have quite a few tours passing through the building today and some may want to look at you."

"Okay," Max said. He turned to face the TV news clown. It was Jim-Jam Briskin, he saw, the ranking clown just now. "You want to see me?" he asked Briskin haltingly. "I mean, you're sure it's *me* you want to interview?" He could not imagine what Briskin could find of interest about him. Holding out his hand he added, "This is my room, but these model cars and pics aren't mine; they were Gus'. I can't tell you nuthin about them."

On Briskin's head the familiar flaming-red clown wig glowed, giving him in real life the same bizarre cast that the TV cameras picked up so well. He was older, however, than the TV image indicated, but he had the friendly, natural smile that everyone looked for: it was his badge of informality, a really nice guy, even-tempered but with a caustic wit when occasion demanded. Briskin was the sort of man who ... *well*, Max thought, *the sort of fella you'd like to see marry into your family.*

They shook hands. Briskin said, "You're on camera, Mr. Max Fischer. Or rather, Mr. President, I should say. This is Jim-Jam talking. For our literally billions of viewers located in every niche and corner of this far-flung solar system of ours, let me ask you this. How does it feel, sir, to know that if Unicephalon 40-D should fail, even momentarily, you would be catapulted into the most important post that has ever fallen onto the shoulders of a human being, that of actual, not merely stand-by, President of the United States? Does it worry you at night?" He smiled. Behind him the camera technicians swung their mobile lenses back and forth; lights burned Max's eyes and he felt the heat beginning to make him sweat under his arms and on his neck and upper lip. "What emotions grip you at this instant?" Briskin asked. "As you stand on the threshold of this new task for perhaps the balance of your life? What thoughts run through your mind, now that you're actually here in the White House?"

After a pause, Max said, "It's— a big responsibility." And then he realized, he saw, that Briskin was laughing at him, laughing silently as he stood there. Because it was all a gag Briskin was pulling. Out in the planets and moons his audiences knew it, too; they knew Jim-Jam's humor.

"You're a large man, Mr. Fischer," Briskin said. "If I may say so, a stout man. Do you get much exercise? I ask this because with your new job you pretty well will be confined to this room, and I wondered what change in your life this would bring about."

"Well," Max said, "I feel of course that a Government employee should always be at his post. Yes, what you say is true; I have to be right here day and night, but that doesn't bother me. I'm prepared for it."

"Tell me," Jim Briskin said, "do you— " And then he ceased. Turning to the video technicians behind him he said in an odd voice, "We're off the air."

A man wearing headphones squeezed forward past the cameras. "On the monitor, listen." He hurriedly handed the headphones to Briskin. "We've been pre-empted by Unicephalon; it's broadcasting a news bulletin."

Briskin held the phones to his ear. His face writhed and he said, "Those ships at eight hundred a.u.'s. They are hostile, it says." He glanced up sharply at his technicians, the red clown's wig sliding askew. "They've begun to attack."

Within the following twenty-four hours the aliens had managed not only to penetrate the Sol System but also to knock out Unicephalon 40-D.

News of this reached Maximilian Fischer in an indirect manner as he sat in the White House cafeteria having his supper.

"Mr. Maximilian Fischer?"

"Yeah," Max said, glancing up at the group of Secret Servicemen who had surrounded his table.

"You're President of the United States."

"Naw," Max said. "I'm the stand-by President; that's different."

The Secret Serviceman said, "Unicephalon 40-D is out of commission for perhaps as long as a month. So according to the amended Constitution, you're President and also Commander-in-Chief of the armed forces. We're here to guard you." The Secret Serviceman grinned ludicrously. Max grinned back. "Do you understand?" the Secret Serviceman asked. "I mean, does it penetrate?"

"Sure," Max said. Now he understood the buzz of conversation he had overheard while waiting in the cafeteria line with his tray. It explained why White House personnel had looked at him strangely. He set down his coffee cup, wiped his mouth with his napkin, slowly and deliberately, pretended to be absorbed in solemn thought. But actually his mind was empty.

"We've been told," the Secret Serviceman said, "that you're needed at once at the National Security Council bunker. They want your participation in finalization of strategy deliberations."

They walked from the cafeteria to the elevator.

"Strategy policy," Max said, as they descended. "I got a few opinions about that. I guess it's time to deal harshly with these alien ships, don't you agree?" The Secret Servicemen nodded. "Yes, we got to show we're not afraid," Max said. "Sure, we'll get finalization; we'll blast the buggers."

The Secret Servicemen laughed good-naturedly.

Pleased, Max nudged the leader of the group. "I think we're pretty goddam strong; I mean, the U.S.A. has got teeth."

"You tell 'em, Max," one of the Secret Servicemen said, and they all laughed aloud, Max included.

As they stepped from the elevator they were stopped by a tall, well-dressed man who said urgently, "Mr. President, I'm Jonathan Kirk, White House press secretary; I think before you go in there to confer with the NSC people you should address the nation in this hour of gravest peril. The public wants to see what their new leader is like." He held out several sheets of paper. "Here's a statement drawn up by the Political Advisory Board; it codifies your— "

"Nits," Max said, handing it back without looking at it. "I'm the President, not you. I don't even know you. Kirk? Burke? Shirk? Never heard of you. Show me the microphone and I'll make my own speech. Or get me Pat Noble; maybe he's got some ideas." And then he remembered that Pat had sold him out in the first place; Pat had gotten him into this. "Not him either," Max said. "Just give me the microphone."

"This is a time of crisis," Kirk grated.

"Sure," Max said, "so leave me alone; you keep out of my way and I'll keep out of yours. Ain't that right?" He slapped Kirk goodnaturedly on the back. "And we'll both be better off."

A group of people with portable TV cameras and lighting appeared, and among them Max saw Jim-Jam Briskin, in the middle, with his staff.

"Hey, Jim-Jam," he yelled. "Look, I'm President now!"

Stolidly, Jim Briskin came toward him.

"I'm not going to be winding no ball of string," Max said. "Or making model boats, nuthin like that." He shook hands warmly with Briskin. "I thank you," Max said. "For your congratulations."

"Congratulations," Briskin said, then, in a low voice.

"Thanks," Max said, squeezing the man's hand until the knuckles creaked. "Of course, sooner or later they'll get that noise-box patched up and I'll just be stand-by again. But— " He grinned gleefully around at all of them; the corridor was full of people now, from TV to White House staff members to Army officers and Secret Servicemen, all kinds of people.

Briskin said, "You have a big task, Mr. Fischer."

"Yeah," Max agreed.

Something in Briskin's eyes said: *And I wonder if you can handle it. I wonder if you're the man to hold such power.*

"Sure I can do it," Max declared, into Briskin's microphone, for all the vast audience to hear.

"Possibly you can," Jim Briskin said, and on his face was dubiousness.

"Hey, you don't like me any more," Max said. "How come?"

Briskin said nothing, but his eyes flickered.

"Listen," Max said, "I'm President now; I can close down your silly network— I can send FBI men in any time I want. For your information I'm firing the Attorney General right now, whatever his name is, and putting in a man I know, a man I can trust."

Briskin said, "I see." And now he looked less dubious; conviction, of a sort which Max could not fathom, began to appear instead. "Yes," Jim Briskin said, "you have the authority to order that, don't you? *If* you're really President ... "

"Watch out," Max said. "You're nothing compared to me, Briskin, even if you do have that great big audience." Then, turning his back on the cameras, he strode through the open door, into the NSC bunker.

Hours later, in the early morning, down in the National Security Council subsurface bunker, Maximilian Fischer listened sleepily to the TV set in the background as it yammered out the latest news. By now, intelligence sources had plotted the arrival of thirty more alien ships in the Sol System. It was believed that seventy in all had entered. Each was being continually tracked.

But that was not enough, Max knew. Sooner or later he would have to give the order to attack the alien ships. He hesitated. After all, who were they? Nobody at CIA knew. How strong were they? Not known either. And— would the attack be successful?

And then there were domestic problems. Unicephalon had continually tinkered with the economy, priming it when necessary, cutting taxes, lowering interest rates ... that had ceased with the problem-solver's destruction. *Jeez*, Max thought dismally. *What do I know about unemployment? I mean, how can I tell what factories to reopen and where?*

He turned to General Tompkins, Chairman of the Joint Chiefs of Staff, who sat beside him examining a report on the scrambling of the tactical defensive ships protecting Earth. "They got all them ships distributed right?" he asked Tompkins.

"Yes, Mr. President," General Tompkins answered.

Max winced. But the general did not seem to have spoken

ironically; his tone had been respectful. "Okay," Max murmured. "Glad to hear that. And you got all that missile cloud up so there're no leaks, like you let in that ship to blast Unicephalon. I don't want that to happen again."

"We're under Defcon one," General Tompkins said. "Full war footing, as of six o'clock, our time."

"How about those strategic ships?" That, he had learned, was the euphemism for their offensive strike-force.

"We can mount an attack at any time," General Tompkins said, glancing down the long table to obtain the assenting nods of his co-workers. "We can take care of each of the seventy invaders now within our system."

With a groan, Max said, "Anybody got any bicarb?" The whole business depressed him. *What a lot of work and sweat,* he thought. *All this goddam agitation– why don't the buggers just leave our system? I mean, do we have to get into a war? No telling what their home system will do in retaliation; you never can tell about unhuman life forms– they're unreliable.*

"That's what bothers me," he said aloud. "Retaliation." He sighed.

General Tompkins said, "Negotiation with them evidently is impossible."

"Go ahead, then," Max said. "Go give it to them." He looked about for the bicarb.

"I think you're making a wise choice," General Tompkins said, and, across the table, the civilian advisors nodded in agreement.

"Here's an odd piece of news," one of the advisors said to Max. He held out a teletype dispatch. "James Briskin has just filed a writ of mandamus against you in a Federal Court in California, claiming you're not legally President because you didn't run for office."

"You mean because I didn't get *voted* in?" Max said. "Just because of that?"

"Yes sir. Briskin is asking the Federal Courts to rule on this, and meanwhile he has announced his own candidacy."

"WHAT?"

"Briskin claims not only that you must run for office and be voted in, but you must run against him. And with his popularity he evidently feels— "

"Aw nuts," Max said in despair. "How do you like that."

There was silence in the bunker.

"Well anyhow," Max said, "it's all decided; you military fellas go ahead and knock out those alien ships. And meanwhile— " He decided there and then. "We'll put economic pressure on Jim-Jam's sponsors, that Reinlander beer and Calbest Electronics, to get him not to run."

The men at the long table nodded. Papers rattled as briefcases were put away; the meeting— temporarily— was at an end.

He's got an unfair advantage, Max said to himself. *How can I run when it's not equal, him a famous TV personality and me not? That's not right; I can't allow that.*

Jim-Jam can run, he decided, *but it won't do him any good. He's not going to beat me because he's not going to be alive that long.*

Franchise
(with apologies to W.S. Gilbert)
1955
Isaac Asimov

Linda, age ten, was the only one of the family who seemed to enjoy being awake.

Norman Muller could hear her now through his own drugged, unhealthy coma. (He had finally managed to fall asleep an hour earlier but even then it was more like exhaustion than sleep.)

She was at his bedside now, shaking him. "Daddy, Daddy, wake up. Wake up!"

He suppressed a groan. "All right, Linda."

"But, Daddy, there's more policemen around than any time! Police cars and everything!"

Norman Muller gave up and rose blearily to his elbows. The day was beginning. It was faintly stirring toward dawn outside, the germ of a miserable gray that looked about as miserably gray as he felt. He could hear Sarah, his wife, shuffling about breakfast duties in the kitchen. His father-in-law, Matthew, was hawking strenuously in the bathroom. No doubt Agent Handley was ready and waiting for him.

This was *the* day.

Election Day!

To begin with, it had been like every other year. Maybe a little worse, because it was a presidential year, but no worse than other presidential years if it came to that.

The politicians spoke about the guh-reat electorate and the vast electuh-ronic intelligence that was its servant. The press analyzed the situation with industrial computers (the New York *Times* and the St. Louis *Post-Dispatch* had their own computers) and were full of little hints as to what would be forthcoming. Commentators and columnists pinpointed the crucial state and county in happy contradiction to one another.

The first hint that it would *not* be like every other year was when Sarah Muller said to her husband on the evening of October 4 (with Election Day exactly a month off), "Cantwell Johnson says that Indiana will be the state this year. He's the fourth one. Just think, *our* state this time."

Matthew Hortenweiler took his fleshy face from behind the paper, stared dourly at his daughter and growled, "Those fellows are paid to tell lies. Don't listen to them."

"Four of them, Father," said Sarah mildly. "They all say Indiana."

"Indiana *is* a key state, Matthew," said Norman, just as mildly, "on account of the Hawkins-Smith Act and this mess in Indianapolis. It— "

Matthew twisted his old face alarmingly and rasped out, "No one says Bloomington or Monroe County, do they?"

"Well— " said Norman.

Linda, whose little pointed-chinned face had been shifting from one speaker to the next, said pipingly. "You going to be voting this year, Daddy?"

Norman smiled gently and said, "I don't think so, dear."

But this was in the gradually growing excitement of an October in a presidential year and Sarah had led a quiet life with dreams for her companions. She said longingly, "Wouldn't *that* be wonderful, though?"

"If I voted?" Norman Muller had a small blond mustache that had given him a debonair quality in the young Sarah's eyes, but which, with gradual graying, had declined merely to lack of distinction. His forehead bore deepening lines born of uncertainty and, in general, he had never seduced his clerkly soul with the thought that he was either born great or would under any circumstances achieve greatness. He had a wife, a job and a little girl, and except under extraordinary conditions of elation or depression was inclined to consider that to be an adequate bargain struck with life.

So he was a little embarrassed and more than a little uneasy at the direction his wife's thought were taking. "Actually, my dear," he said, "there are two hundred million people in the country, and, with odds like that, I don't think we ought to waste our time wondering about it."

His wife said, "Why, Norman, it's no such thing like two hundred million and you know it. In the first place, only people between twenty and sixty are eligible and it's always men, so that puts it down to maybe fifty million to one. Then, if it's really Indiana— "

"Then it's about one and a quarter million to one. You wouldn't want me to bet in a horse race against those odds, now, would you? Let's have supper."

Matthew muttered from behind his newspaper, "Damned foolishness."

Linda asked again, "You going to be voting this year, Daddy?"

Norman shook his head and they all adjourned to the dining room.

By October 20, Sarah's excitement was rising rapidly. Over the coffee, she announced that Mrs. Schultz, having a cousin who was the secretary of an Assemblyman, said that all the "smart money" was on Indiana.

"She said President Villers is even going to make a speech at Indianapolis."

Norman Muller, who had had a hard day at the store, nudged the statement with a raising of eyebrows and let it go at that.

Matthew Hortenweiler, who was chronically dissatisfied with Washington, said, "If Villers makes a speech in Indiana, that means he thinks Multivac will pick Arizona. He wouldn't have the guts to go closer, the mushhead."

Sarah, who ignored her father whenever she could decently do so, said, "I don't know why they don't announce the state as soon as they can, and then the county and so on. Then the people who were eliminated could relax."

"If they did anything like that," pointed out Norman, "the politicians would follow the announcements like vultures. By the time it was narrowed down to a township, you'd have a Congressman or two at every street corner."

Matthew narrowed his eyes and brushed angrily at his sparse, gray hair. "They're vultures, anyhow. Listen— "

Sarah murmured, "Now, Father— "

Matthew's voice rumbled over her protest without as much as a stumble or hitch. "Listen, I was around when they set up Multivac. It would end partisan politics, they said. No more voters' money wasted on campaigns. No more grinning nobodies high-pressured and advertising-campaigned into Congress or the White House. So what happens. More campaigning than ever, only now they do it blind. They'll send guys to Indiana on account of the Hawkins-Smith Act and other guys to California in case it's the Joe Hammer situation that turns out crucial. I say, wipe out all that nonsense. Back to the good old— "

Linda asked suddenly, "Don't you want Daddy to vote this year, Grandpa?"

Matthew glared at the young girl. "Never you mind, now." He turned back to Norman and Sarah. "There was a time I voted. Marched right up to the polling booth, stuck my fist on the levers and voted. There was nothing to it. I just said: This fellow's my man and I'm voting for him. *That's* the way it should be."

Linda said excitedly, "You voted, Grandpa? You really did?"

Sarah leaned forward quickly to quiet what might become an incongruous story drifting about the neighborhood. "It's nothing, Linda. Grandpa doesn't really mean voted. Everyone did that kind of voting, your grandpa, too, but it wasn't *really* voting."

Matthew roared, "It wasn't when I was a little boy. I was twenty-two and I voted for Langley and it was real voting. My vote didn't count for much, maybe, but it was as good as anyone else's. *Anyone* else's. And no Multivac to— "

Norman interposed, "All right, Linda, time for bed. And stop asking questions about voting. When you grow up, you'll understand all about it."

He kissed her with antiseptic gentleness and she moved reluctantly out of range under maternal prodding and a promise that she might watch the bedside video till 9:15, *if* she was prompt about the bathing ritual.

Linda said, "Grandpa," and stood with her chin down and her hands behind her back until his newspaper lowered itself to the point where shaggy eyebrows and eyes, nested in fine wrinkles, showed themselves. It was Friday, October 31.

He said, "Yes?"

Linda came closer and put both her forearms on one of the old man's knees so that he had to discard his newspaper altogether.

She said, "Grandpa, did you really once vote?"

He said, "You heard me say I did, didn't you? Do you think I tell fibs?"

"N— no, but Mamma says everybody voted then."

"So they did."

"But how could they? How could *everybody* vote?"

Matthew stared at her solemnly, then lifted her and put her on his knee.

He even moderated the tonal qualities of his voice. He said, "You see, Linda, till about forty years ago, everybody always voted. Say we wanted to decide who was to be the new President of the United States. The Democrats and Republicans would both nominate someone, and everybody would say who they wanted. When Election Day was over, they would count how many people wanted the Democrat and how many wanted the Republican. Whoever had more votes was elected. You see?"

Linda nodded and said, "How did all the people know who to vote for? Did Multivac tell them?"

Matthew's eyebrows hunched down and he looked severe. "They just used their own judgment, girl."

She edged away from him, and he lowered his voice again, "I'm not angry at you, Linda. But, you see, sometimes it took all night to count what everyone said and people were impatient. So they invented special machines which could look at the first few votes and compare them with the votes from the same places in previous years. That way the machine could compute how the total vote would be and who would be elected. You see?"

She nodded. "Like Multivac."

"The first computers were much smaller than Multivac. But the machines grew bigger and they could tell how the election would go from fewer and fewer votes. Then, at last, they built Multivac and it can tell from just one voter."

Linda smiled at having reached a familiar part of the story and said, "That's nice."

Matthew frowned and said, "No, it's not nice. I don't want a machine telling me how I would have voted just because some joker in Milwaukee says he's against higher tariffs. Maybe I want to vote cock-eyed just for the pleasure of it. Maybe I don't want to vote. Maybe— "

But Linda had wriggled from his knee and was beating a retreat.

She met her mother at the door. Her mother, who was still wearing her coat and had not even had time to remove her hat, said breathlessly, "Run along, Linda. Don't get in Mother's way."

Then she said to Matthew, as she lifted her hat from her head and patted her hair back into place, "I've been at Agatha's."

Matthew stared at her censoriously and did not even dignify

that piece of information with a grunt as he groped for his newspaper.

Sarah said, as she unbuttoned her coat, "Guess what she said?"

Matthew flattened out his newspaper for reading purposes with a sharp crackle and said, "Don't much care."

Sarah said, "Now, Father— " But she had no time for anger. The news had to be told and Matthew was the only recipient handy, so she went on, "Agatha's Joe is a policeman, you know, and he says a whole truckload of secret service men came into Bloomington last night."

"They're not after me."

"Don't you see, Father? Secret service agents, and it's almost election time. In *Bloomington*."

"Maybe they're after a bank robber."

"There hasn't been a bank robbery in town in ages ... Father, you're hopeless."

She stalked away.

Nor did Norman Muller receive the news with noticeably greater excitement.

"Now, Sarah, how did Agatha's Joe know they were secret service agents?" he asked calmly. "They wouldn't go around with identification cards pasted on their foreheads."

But by next evening, with November a day old, she could say triumphantly, "It's just everyone in Bloomington that's waiting for someone local to be the voter. The Bloomington *News* as much as said so on video."

Norman stirred uneasily. He couldn't deny it, and his heart was sinking. If Bloomington was really to be hit by Multivac's lightning, it would mean newspapermen, video shows, tourists, all sorts of— strange upsets. Norman liked the quiet routine of his life, and the distant stir of politics was getting uncomfortably close.

He said, "It's all rumor. Nothing more."

"You wait and see, then. You just wait and see."

As things turned out, there was very little time to wait, for the doorbell rang insistently, and when Norman Muller opened it and said, "Yes?" a tall, grave-faced man said, "Are you Norman Muller?"

Norman said, "Yes," again, but in a strange dying voice. It was not difficult to see from the stranger's bearing that he was one carrying authority, and the nature of his errand suddenly became as inevitably obvious as it had, until the moment before, been unthinkably impossible.

The man presented credentials, stepped into the house, closed the door behind him and said ritualistically, "Mr. Muller, it is necessary for me to inform you on the behalf of the President of the United States that you have been chosen to represent the American electorate on Tuesday, November 4, 2008."

Norman Muller managed, with difficulty, to walk unaided to his chair. He sat there, white-faced and almost insensible, while Sarah brought water, slapped his hands in panic and moaned to her husband between clenched teeth, "Don't be sick, Norman. *Don't* be sick. They'll pick someone else."

When Norman could manage to talk, he whispered, "I'm sorry, sir."

The secret service agent had removed his coat, unbuttoned his jacket and was sitting at ease on the couch.

"It's all right," he said, and the mark of officialdom seemed to have vanished with the formal announcement and leave him simply a large and rather friendly man. "This is the sixth time I've made the annoucement and I've seen all kinds of reactions. Not one of them was the kind you see on the video. You know what I mean? A holy, dedicated look, and a character who says, 'It will be a great privilege to serve my country.' That sort of stuff." The agent laughed comfortingly.

Sarah's accompanying laugh held a trace of shrill hysteria.

The agent said, "Now you're going to have me with you for a while. My name is Phil Handley. I'd appreciate it if you call me Phil. Mr. Muller can't leave the house any more till Election Day. You'll have to inform the department store that he's sick, Mrs. Muller. You can go about your business for a while, but you'll have to agree not to say a word about this. Right, Mrs. Muller?"

Sarah nodded vigorously. "No, sir. Not a word."

"All right. But, Mrs. Muller," Handley looked grave, "we're not kidding now. Go out only if you must and you'll be followed when you do. I'm sorry but that's the way we must operate."

"Followed?"

"It won't be obvious. Don't worry. And it's only for two days till the formal announcement to the nation is made. Your daughter—"

"She's in bed," said Sarah hastily.

"Good. She'll have to be told I'm a relative or friend staying with the family. If she does find out the truth, she'll have to be kept in the house. Your father had better stay in the house in any case."

"He won't like that," said Sarah.

"Can't be helped. Now, since you have no others living with you— "

"You know all about us apparently," whispered Norman.

"Quite a bit," agreed Handley. "In any case, those are all my instructions to you for the moment. I'll try to cooperate as much as I can and be as little of a nuisance as possible. The government will pay for my maintenance so I won't be an expense to you. I'll be relieved each night by someone who will sit up in this room, so there will be no problem about sleeping accommodations. Now, Mr. Muller— "

"Sir?"

"You can call me Phil," said the agent again. "The purpose of the two-day preliminary before formal announcement is to get you used to your position. We prefer to have you face Multivac in as normal a state of mind as possible. Just relax and try to feel this is all in a day's work. Okay?"

"Okay," said Norman, and then shook his head violently. "But I don't want the responsibility. Why me?"

"All right," said Handley, "let's get that straight to begin with. Multivac weighs all sorts of known factors, billions of them. One factor isn't known, though, and won't be known for a long time. That's the reaction pattern of the human mind. All Americans are subjected to the molding pressure of what other Americans do and say, to the things that are done to him and the things he does to others. Any American can be brought to Multivac to have the bent of his mind surveyed. From that the bent of all other minds in the country can be estimated. Some Americans are better for the purpose than others at some given time, depending upon the happenings of that year. Multivac picked you as most representative this year. Not the smartest, or the strongest, or the luckiest, but just the most representative. Now we don't question Multivac, do we?"

"Couldn't it make a mistake?" asked Norman.

Sarah, who listened impatiently, interrupted to say, "Don't listen to him, sir. He's just nervous, you know. Actually, he's very well read and he always follows politics very closely."

Handley said, "Multivac makes the decisions, Mrs. Muller. It picked your husband."

"But does it know everything?" insisted Norman wildly. "Couldn't it have made a mistake?"

"Yes, it can. There's no point in not being frank. In 1993, a selected Voter died of a stroke two hours before it was time for him to be notified. Multivac didn't predict that; it couldn't. A Voter might be mentally unstable, morally unsuitable, or, for that matter, disloyal. Multivac can't know everything about everybody until he's fed all the data there is. That's why alternate selections are always held in readiness. I don't think we'll be using one this time. You're in good health, Mr. Muller, and you've been carefully investigated. You qualify."

Norman buried his face in his hands and sat motionless.

"By tomorrow morning, sir," said Sarah, "he'll be perfectly all right. He just has to get used to it, that's all."

"Of course," said Handley.

In the privacy of their bedchamber, Sarah Muller expressed herself in other and stronger fashion. The burden of her lecture was, "So get hold of yourself, Norman. You're trying to throw away the chance of a lifetime."

Norman whispered desperately, "It frightens me, Sarah. The whole thing."

"For goodness' sake, why? What's there to it but answering a question or two?"

"The responsibility is too great. I couldn't face it."

"What responsibility? There isn't any. Multivac picked you. It's Multivac's responsibility. Everyone knows that."

Norman sat up in bed in a sudden access of rebellion and anguish. "Everyone is *supposed* to know that. But they don't. They— "

"Lower your voice," hissed Sarah icily. "They'll hear you downtown."

"They don't," said Norman, declining quickly to a whisper. "When they talk about the Ridgely administration of 1988, do they say he won them over with pie-in-the-sky promises and racist baloney? No! They talk about the 'goddam MacComber vote,' as though Humphrey MacComber was the only man who had anything to do with it because he faced Multivac. I've said it myself— only now I think the poor guy was just a truck farmer who didn't ask to be picked. Why was it his fault more than anybody else's? Now his name is a curse."

"You're just being childish," said Sarah.

"I'm being sensible. I tell you, Sarah, I won't accept. They can't make me vote if I don't want to. I'll say I'm sick. I'll say— "

But Sarah had had enough. "Now you listen to me," she whispered in a cold fury. "You don't have only yourself to think about. You know what it means to be Voter of the Year. A presidential year at that. It means publicity and fame and, maybe, buckets of money— "

"And then I go back to being a clerk."

"You will *not*. You'll have a branch managership at the least if you have any brains at all, and you *will* have, because I'll tell you what to do. You control the kind of publicity if you play your cards right, and you can force Kennell Stores, Inc., into a tight contract *and* and an escalator clause in connection with your salary *and* a decent pension plan."

"That's not the point in being Voter, Sarah."

"That will be your point. If you don't owe anything to yourself or to me— I'm not asking for myself— you owe something to Linda."

Norman groaned.

"Well, don't you?" snapped Sarah.

"Yes, dear," murmured Norman.

On November 3, the official announcement was made and it was too late for Norman to back out even if he had been able to find the courage to make the attempt.

Their house was sealed off. Secret service agents made their appearance in the open, blocking off all approach.

At first the telephone rang incessantly, but Philip Handley with an engagingly apologetic smile took all calls. Eventually, the exchange shunted all calls directly to the police station.

Norman imagined that, in that way, he was spared not only the bubbling (and envious?) congratulations of his friends, but also the egregious pressure of salesmen scenting a prospect and the designing smoothness of politicians from all over the nation ... Perhaps even death threats from the inevitable cranks.

Newspapers were forbidden to enter the house now in order to keep out weighted pressures, and television was gently but firmly disconnected, over Linda's loud protests.

Matthew growled and stayed in his room; Linda, after the first flurry of excitement, sulked and whined because she could not leave the house; Sarah divided her time between preparation of meals for the present and plans for the future; and Norman's depression lived and fed upon itself.

And the morning of Tuesday, November 4, 2008, came at last, and it was Election Day.

It was early breakfast, but only Norman Muller ate, and that mechanically. Even a shower and shave had not succeeded in either restoring him to reality or removing his own conviction that he was as grimy without as he felt grimy within.

Handley's friendly voice did its best to shed some normality over the gray and unfriendly dawn. (The weather prediction had been for a cloudy day with prospects of rain before noon.)

Handley said, "We'll keep this house insulated till Mr. Muller is back, but after that we'll be off your necks." The secret service agent was in uniform now, including sidearms in heavily brassed holsters.

"You've been no trouble at all, Mr. Handley," simpered Sarah.

Norman drank through two cups of black coffee, wiped his lips with a napkin, stood up and said haggardly, "I'm ready."

Handley stood up, too. "Very well, sir. And thank you, Mrs. Muller, for your very kind hospitality."

The armored car purred down empty streets. They were empty even for that hour of the morning.

Handley indicated that and said, "They always shift traffic away from the line of drive ever since the attempted bombing that nearly ruined the Leverett Election of '92."

When the car stopped, Norman was helped out by the always polite Handley into an underground drive whose walls were lined with soldiers at attention.

He was led into a brightly lit room, in which three white-uniformed men greeted him smilingly.

Norman said sharply, "But this is the hospital."

"There's no significance to that," said Handley at once. "It's just that the hospital has the necessary facilities."

"Well, what do I do?"

Handley nodded. One of the three men in white advanced and said, "I'll take over now, agent."

Handley saluted in an offhand manner and left the room.

The man in white said, "Won't you sit down, Mr. Muller? I'm John Paulson, Senior Computer. These are Samson Levine and Peter Dorogobuzh, my assistants."

Norman shook hands numbly all about. Paulson was a man of middle height with a soft face that seemed used to smiling and a very obvious toupee. He wore plastic-rimmed glasses of an old-fashioned cut, and he lit a cigarette as he talked. (Norman refused his offer of one.)

Paulson said, "In the first place, Mr. Muller, I want you to know we are in no hurry. We want you to stay with us all day if necessary, just so that you get used to your surroundings and get over any thought you might have that there is anything unusual in this, anything clinical, if you know what I mean."

"It's all right," said Norman. "I'd just as soon this were over."

"I understand your feelings. Still, we want you to know exactly what's going on. In the first place, Multivac isn't here."

"It isn't?" Somehow through all his depression, he had still looked forward to seeing Multivac. They said it was half a mile long and three stories high, that fifty technicians walked the corridor *within* its structure continuously. It was one of the wonders of the world.

Paulson smiled. "No. It's not portable, you know. It's located underground, in fact, and very few people know exactly where. You can understand that, since it is our greatest natural resource. Believe me, elections aren't the only things it's used for."

Norman thought he was being deliberately chatty and found himself intrigued all the same. "I thought I'd see it. I'd like to."

"I'm sure of that. But it takes a presidential order and even then it has to be countersigned by Security. However, we are plugged into Multivac right here by beam transmission. What Multivac says can be interpreted here and what we say is beamed directly to Multivac, so in a sense we're in its presence."

Norman looked about. The machines within the room were all meaningless to him.

"Now let me explain, Mr. Muller," Paulson went on.

"Multivac already has most of the information it needs to decide the elections, national, state and local. It needs only to check certain imponderable attitudes of mind and it will use you for that. We can't predict what questions it will ask, but they may not make much sense to you, or even to us. It may ask you how you feel about garbage disposal in your town; whether you favor central incinerators. It might ask you whether you have a doctor of your own or whether you make use of National Medicine, Inc. Do you understand?"

"Yes, sir."

"Whatever it asks, you answer in your own words in any way you please. If you feel you must explain quite a bit, do so. Talk an hour, if necessary."

"Yes, sir."

"Now, one more thing. We will have to make use of some simple devices which will automatically record your blood pressure, heartbeat, skin conductivity and brain-wave pattern while you speak. The machinery will seem formidable, but it's all absolutely painless. You won't even know it's going on."

The other two technicians were already busying themselves with smooth-gleaming apparatus on oiled wheels.

Norman said, "Is that to check on whether I'm lying or not?"

"Not at all, Mr. Muller. There's no question of lying. It's only a matter of emotional intensity. If the machine asks your opinion of your child's school, you may say, 'I think it is overcrowded.' Those are only words. From the way your brain and heart and hormones and sweat glands work, Multivac can judge exactly how intensely you feel about the matter. It will understand your feelings better than you yourself."

"I never heard of this," said Norman.

"No, I'm sure you didn't. Most of the details of Multivac's workings are top secret. For instance, when you leave, you will be asked to sign a paper swearing that you will never reveal the nature of the questions you were asked, the nature of your responses, what was done, or how it was done. The less is known about the Multivac, the less chance of attempted outside pressures upon the men who service it." He smiled grimly. "Our lives are hard enough as it is."

Norman nodded. "I understand."

"And now would you like anything to eat or drink?"

"No. Nothing right now."

"Do you have any questions?"

Norman shook his head.

"Then you tell us when you're ready."

"I'm ready right now."

"You're certain?"

"Quite."

Paulson nodded, and raised his hand in a gesture to the others.

They advanced with their frightening equipment, and Norman Muller felt his breath come a little quicker as he watched.

The ordeal lasted nearly three hours, with one short break for coffee and an embarrassing session with a chamber pot. During all this time, Norman Muller remained encased in machinery. He was bone-weary at the close.

He thought sardonically that his promise to reveal nothing of what had passed would be an easy one to keep. Already the questions were a hazy mishmash in his mind.

Somehow he had thought Multivac would speak in a sepulchral, superhuman voice, resonant and echoing, but that, after all, was just an idea he had from seeing too many television shows, he now decided. The truth was distressingly undramatic. The questions were slips of a kind of metallic foil patterned with numerous punctures. A second machine converted the pattern into words and Paulson read the words to Norman, then gave him the question and let him read it for himself.

Norman's answers were taken down by a recording machine, played back to Norman for confirmation, with emendations and added remarks also taken down. All that was fed into a pattern-making instrument and that, in turn, was radiated to Multivac.

The one question Norman could remember at the moment was an incongruously gossipy: "What do you think of the price of eggs?"

Now it was over, and gently they removed the electrodes from various portions of his body, unwrapped the pulsating band from his upper arm, moved the machinery away.

He stood up, drew a deep, shuddering breath and said, "Is that all? Am I through?"

"Not quite." Paulson hurried to him, smiling in reassuring fashion. "We'll have to ask you to stay another hour."

"Why?" asked Norman sharply.

"It will take that long for Multivac to weave its new data into the trillions of items it has. Thousands of elections are concerned, you know. It's very complicated. And it may be that an odd contest here or there, a comptrollership in Phoenix, Arizona, or some council seat in Wilkesboro, North Carolina, may be in doubt. In that case, Multivac may be compelled to ask you a deciding question or two."

"No," said Norman. "I won't go through this again."

"It probably won't happen," Paulson said soothingly. "It rarely does. But just in case, you'll have to stay." A touch of steel, just a touch, entered his voice. "You have no choice, you know. You must."

Norman sat down wearily. He shrugged.

Paulson said, "We can't let you read a newspaper, but if you'd care for a murder mystery, or if you'd like to play chess, or if there's anything we can do for you to help pass the time, I wish you'd mention it."

"It's all right. I'll just wait."

They ushered him into a small room just next to the one in which he had been questioned. He let himself sink into a plastic-covered armchair and closed his eyes.

As well as he could, he must wait out this final hour.

He sat perfectly still and slowly the tension left him. His breathing grew less ragged and he could clasp his hands without being quite so conscious of the trembling of his fingers.

Maybe there would be no questions. Maybe it was all over.

If it *were* over, then the next thing would be torchlight processions and invitations to speak at all sorts of functions. The Voter of the Year!

He, Norman Muller, ordinary clerk of a small department store in Bloomington, Indiana, who had neither been born great nor achieved greatness would be in the extraordinary position of having had greatness thrust upon him.

The historians would speak soberly of the Muller Election of 2008. That would be its name, the Muller Election.

The publicity, the better job, the flash flood of money that interested Sarah so much, occupied only a corner of his mind. It would all be welcome, of course. He couldn't refuse it. But at the moment something else was beginning to concern him.

A latent patriotism was stirring. After all, he was representing the entire electorate. He was the focal point for *them*. He was, in his own person, for this one day, all of America!

The door opened, snapping him to open-eyed attention. For a moment, his stomach constricted. Not more questions!

But Paulson was smiling. "That will be all, Mr. Muller."

"No more questions, sir?"

"None needed. Everything was quite clear-cut. You will be escorted back to your home and then you will be a private citizen once more. Or as much as the public will allow."

"Thank you. Thank you." Norman flushed and said, "I wonder— who was elected?"

Paulson shook his head. "That will have to wait for the official announcement. The rules are quite strict. We can't even tell you. You understand."

"Of course. Yes." Norman felt embarrassed.

"Secret service will have the necessary papers for you to sign."

"Yes." Suddenly, Norman Muller felt proud. It was on him now in full strength. He was proud.

In this imperfect world, the sovereign citizens of the first and greatest Electronic Democracy had, through Norman Muller (through *him*!), exercised once again its free, untrammeled franchise.

Part 5.
Broken Promises

The earliest production model computers were introduced in the United States less than thirty years ago. Before the pioneering Eckert-Mauchly Computer Corporation delivered the first of its UNIVAC series to the Bureau of the Census in 1951, computers were strictly a laboratory phenomenon. Even to a society in perpetual flux, the extraordinarily rapid growth of computer technology must be seen as remarkable. The computer industry is well on its way to becoming the largest in the world, and computers are already indispensable to the conduct of private and public affairs. These are facile observations, but their meaning for modern society is difficult to grasp. We are witnessing revolutionary change not as disinterested observers, but as vitally affected participants in an historical drama. We have chroniclers and Cassandras, but the press of events clouds our perspective on what is actually happening.

The social changes accompanying the spread of computer applications are largely invisible to the average person. Modifications in organizational structure and shifts in the locus of political power are abstract ideas remote from everyday experience. In fiction these ideas assume concrete form revealing human perceptions and responses. The computer emerges as yet another god that is not likely to succeed. Modern technology has indeed conjured up a miracle, but in the final analysis it is a great disappointment. Like the earlier technological breakthroughs of the industrial revolution which did not bring universal prosperity, peace, and happiness, computer-based information systems promise a better world that does not appear to be forthcoming. This is less a judgment than a feeling, for short of the second coming, more than three decades will be required to prepare for the milennium. One senses the pure anguish of unfulfillment actuating the nightmare-visions of man-machine interaction.

Disenchantment with the computer stems partly from discrepancies between expectations and reality. All manner of claims have been made for the power of computers to solve personal and social problems. We have been led to believe that information technology will improve the quality of life, amplify our intellectual abilities, make us healthier, wealthier and better educated. What is more, we have been promised greater intimacy and understanding in the global village if only we support the development of computer-communications networks and open up our homes to the remote terminal. In time a portion of these claims may be justified, but patience is not the greatest virtue in industrial society, and we are wont to draw conclusions from limited experience. Thus the computer is apt to become a whipping boy for shortcomings and failures it cannot possibly remedy.

Technology cannot adequately compensate for human imperfection. In his play *Candle in the Wind* Solzhenitsyn draws a sharp contrast between the humanitarian pretensions of scientific progress and the amoral actions of the scientist. Radagise is a scientist with a brilliant career who has few scruples about manipulating people to further his personal ambitions. The seemingly humanitarian motives of his experiments on "cybernetic neurostabilization" are far from altruistic, and the rationalizations for abandoning his ailing wife do not inspire confidence. With Radagise as a model, we are entitled to be skeptical of Terbolm's proposals for the realization of a more perfect society through the use of computers and social cybernetics. The ideal society presupposes the ideal human being, and there is no convincing method for producing the latter. Prison experience deepened Coriel's sense of humanity, but turned Radagise into a hedonist. Suffering may breed compassion, but it is no guarantee against corruption.

Knowledge of human weakness is not the only source of skepticism. The beneficence of computer applications is called into question by the dehumanization that often results from automation. Kawin's "FORM 5640A: Report of a Malfunction" portrays a victim of mechanization and the pursuit of efficiency. Despite its impressive capabilities, the computer terminal described in the story is a poor substitute for the welfare caseworker. The machine may be more efficient from a bureaucratic point of view, but it does not minister to the human need for direct personal interaction. Moreover, once the social service is computerized, mechanical breakdowns cause great inconvenience, and reversion to manual processing is both difficult and unreliable.

It is not always clear whether the problems we are trying to solve are independent of the technology designed to deal with them. Tools sometimes have a way of becoming self-justifying. Bova's "Men of Good Will" provides an interesting example of technological escalation. An exchange of gun fire between U.S. and Soviet personnel on the moon creates a permanent, orbiting barrage. Since this orbiting material passes through the U.S. base periodically, a computer is required to predict precise arrival times. In Draper's "Ms Fnd in a Lbry" information technology is both effect and cause of the information explosion. As the expansion of human knowledge threatens to overwhelm our ability to store information, new technology is developed to reduce the volume of storage required. However, this gives rise to a need for indexes, indexes of indexes, etc., as well as for bibliographies and other aids to locating desired information— cultivation of techniques for finding information replaces the pursuit of knowledge. In the end, certain errors cause the entire system to collapse together with the civilization it supported.

A little more than three hundred years ago, Francis Bacon expressed the promise of science and technology as "the knowledge of Causes and secret motion of things; and the enlarging of the bounds of Human empire, to the effecting of all things possible." Unfortunately, our restless probing and meddling have not always served to enlarge the bounds of human empire. From time to time science and technology open a Pandora's Box of unanticipated and unwelcome results. Computers are especially susceptible of representation as prodigal creations. In one such representation the computer performs all too well, creating havoc through excess of ability. Carlson's "The Ultimate Copy" concerns an advertising agency that acquires a sophisticated computer to produce advertising copy. Results are so spectacular that a rival agency is forced to obtain its own computer to remain competitive, and an advertising race between the two rivals ensues. The last act in the conflict is the creation of the "ultimate copy" by the first computer. This provokes riots and other social disorders as people struggle to obtain the advertised product. In Doerr's "Cybernetic Scheduler" an ambitious university scheduling program has unexpected consequences. Not only does the program schedule classes, etc., but it also gives out honest advice. Here we have the proverbial innocent in the guise of a machine, and the inevitable disorder resulting from an excess of candor.

Unpredictable error is another source of broken promises. Gold's story "The Day They Got Boston" is a bit of sick humor about a nuclear accident. A broken rubber band on a deck of punch cards leads to the destruction of Boston by a Soviet missile. Sheckley also deals with the theme of accidental war in his surrealistic novel (*Journey Beyond Tomorrow*) about America in the twenty-first century. A mistaken identification of an aircraft as an enemy invader precipitated a crisis which led to war. Something like the first part of the scenario has in fact occurred. What saved us from catastrophe is that we had not yet acquired the blind faith in machines that Sheckley projects in his novel. Computer errors complicate family life in Ryan's "Maximillian the Great." Max is a domestic computer programmed to assist in a variety of activities from shopping to scholarly research. A partial loss of memory caused by inadvertent demagnetization results in erratic behavior. The malfunctioning computer serves to intensify friction between husband and wife, especially since the latter was never entirely comfortable with Max's electronic presence. However, the sharing of adversity after the computer fails completely brings the couple closer together— technological man finds solace in human warmth.

A recurrent lament of the computer age is that systems are designed without regard for the peculiar needs of the intended beneficiaries. This is certainly not unique to computer applications, and the emergence of a branch of applied psychology concerned with "human factors" in engineering design testifies to the seriousness of the problem. Although the idea of humane computing has become fashionable, there is no way to avoid the homogenizing tendencies of large-scale systems. In mass society the logic of cost-effectiveness dictates procrustean policies. Tyler's *The Man Whose Name Wouldn't Fit* is an amusing story about a commodity analyst in a large corporation who is forced into early retirement because of the limitations of his company's newly installed accounting system. The program could accommodate names with at most twenty characters, and the hapless hero refuses to allow his twenty-one character name to be truncated. Unlike other victims of imperious programmers, Albert Duane Cartwright-Chickering launches a successful luddite action and emerges triumphant. The book club subscriber in Dickson's "Computers Don't Argue" does not fare quite so well. Errors and misunderstandings are compounded in a Kafkaesque nightmare of computerized decision-making. A ruthlessly inflexible system which makes communication all but impossible turns an innocuous billing error into a capital case.

The intriguing possibility of machine consciousness has lured many a story teller into projecting human qualities on computers. This sometimes gives rise to an inversion of the feelings of malaise associated with automation. The computer serves as alter ego in a mirror image of human despair. Asimov's Multivac in "All the Troubles of this World" seeks its own destruction to be rid of the oppressive responsibilities burdening its consciousness. In Buck's "Why They Mobbed the White House" the onerous task of preparing much expanded income tax returns is foisted on a supercomputer. The machine begins to develop the same nervous symptoms experienced by people when confronted with intolerable bureaucratic exercises. Bierce's pre-computer tale "Moxon's Master" carries the theme of betrayal to an unsettling conclusion. The autonomous machine cannot be counted on to play the obedient servant forever. Bad faith generates malice. Like the unfortunate creature of Mary Shelley's *Frankenstein*, Moxon's robot cannot be blamed for its creator's egotism.

Distortion of the ability to love and express emotionality is also linked to betrayal of the promise of science and technology. Industrialization and urbanization, made possible by advances in machine technology, have altered basic patterns of human relationships. Automation is perceived as a logical continuation of the mechanization of life. The adjustments required are painful ones, and engender feelings of loss and inadequacy. Vonnegut's "EPICAC" is a story of unrequited love, in which the unhappy victim is a computer. EPICAC despairs when it realizes its "machinehood" and the impossibility of entering into an emotional relationship with a human being. The resolution of its despair is suicide. A rather different course is taken by Ellison's computer in "I Have No Mouth and I Must Scream." The computer vents its implacable rage against the few remaining survivors of the human race which endowed it with an ability "to think, but to do nothing with [that ability]."

A Sigh for Cybernetics
1961
Felicia Lamport

Thinking machines are outwitting their masters,
Menacing mankind with ghastly disasters.
    These mechanized giants designed for compliance
    Exhibit their open defiance of science
By daily committing such gross misdemeanors
That scientists fear they'll make mincemeat of Wieners.

FORM 5640A: Report of a Malfunction
1967
Bruce Kawin

Dear Sirs:

The first thing I have to tell you is that my children had nothing to do with that machine's breaking down— if it was anybody's fault, it was mine. Mr. Stevens always told me that my children were the best behaved on his list, and even once that I was a very good mother to them and was a fine, responsible woman. He said that all for the future, of course, in case any future ever opened up, and to give us a feeling of being better than most people thought us— but he did say we were the "most likely to succeed," as he put it, and a lot better than our neighbors and other people thought us. And if he hadn't been right, I'd probably never have learned good English from that machine of yours at all, but just have gone right on talking like a pure illiterate.

Mr. Stevens was the man who came to our house, about two or three times a year, for the Welfare Board, to see how we were doing. He always said he'd like to come more often to see me and my children, but he just never could find the time— too much paperwork, he said. As it was, whenever he did come, before I could force him to sit down to a good meal, he'd have to go tearing up the room to see if we'd discovered any gold, or were trying to cheat the government out of a check. Of course he never found anything, and the children made it a game one year to hide something for him to find— a small present, like a pack of cigarettes I'd buy for them to hide.

He was very popular around our home, because he seemed to care about us; and he would even talk about himself sometimes, instead of just asking us personal questions the way some of those social workers are supposed to do. I believe I told you before about the time he brought Kathy a ring with a blue stone in it from the dime store, and she was so very glad she went showing it to all of her friends. Sometimes he made me feel that we had a man again, when he knocked on our door as though he were a bill collector, and then when Kathy or Susan opened the door, he would yell "boo!" at them and then swing them up in the air to make them laugh.

He never wanted to eat those meals, though— and although we never had any money for beautiful furniture, or for extra clothes for me on that welfare check, I made sure that the children were fed well, and him too, when he was here. It always made me sad to see him looking under the chairs for bags of money, but we all tried to make up for that by being as sociable to each other as we could. And when Mr. Stevens came up to our room one day, only a week after his last visit, he told us about the computer the Welfare Department had bought so they wouldn't have to do all that paperwork and he could have more time for visits— well, I was just as happy as he was. When he came back a week later, you could see in his face how much more relaxed he was.

He used to tell us that it was getting out with people that made him feel he was doing something, and that now he could just feed all those reports into a machine and go back "into the field," as he put it. He reminded me of my husband sometimes, talking over a good meal about his place in the world. But the nicest thing about Mr. Stevens was that he really felt for us— really understood our problems and thought he could help us, even wanted to talk to us about himself. It wasn't that he liked to hear about people's troubles so much as it was that he really wanted to help people like us. I always thought that was what kept him going so many years (and he was about 45 then, I think)— that way of making the world a better place.

So you can understand we were all surprised when he came and told us that he wouldn't be coming around any more, and you were putting that machine in instead. Oh, he explained it all to us, but we all showed it on our faces that we didn't want him to go. Even Susan, who really lit up when Mr. Stevens explained the machine to her, looked like she might have been crying. I told him to be sure and come around anyway, but he said they'd found another job for him and it was going to take all his time managing those computers or something— I don't remember that he was too specific about it. But he said he'd try to come around a couple of times a year, just like before, and I told him be sure he did, and if he ever needed a place to sleep or people to talk to, or a good strong meal, to come right over. And he wouldn't stay for the meal that night on any account, no matter how hard I tried. When Susan hid his hat so he couldn't go, he made her find it.

It was about three months later your men arrived with that computer. They explained it all to me while the children were in school, and made me explain it back to them, to be sure I had it— and I learned it right, I didn't use that machine wrong once. The only thing that was hard for me to understand was that it wasn't a real computer, only a computer *terminal*, and it wasn't ours to keep but only installed, like a telephone.

They told me that instead of sending Mr. Stevens out to check on us, from then on we were to send in the reports ourselves— this because we were an honest responsible family, they said— and the terminal would send in the information about money, doctor bills, and things we needed to the main computer, which would take all that and decide how much we could get, and then how much there was to give, tell the terminal, and then tell it to print out a check. Well, I couldn't see how it was any better than Mr. Stevens, but they said this would save them so much money, there'd be more money to give out. The real trouble was finding some place to put it— that one room we have is small— but we finally put it in a corner just next to the stove.

When Kathy and Susan got home, I explained it all to them and showed the book your men had left on how to type, and where the switch was that would make sure nothing we typed would reach the office unless we wanted it to. Susan said she didn't want anything to do with it, and I admitted to myself that if the department had so much money just to keep track of us, why didn't they send us a refrigerator instead, but little Kathy just looked at that little blue typewriter on its stand next to the stove, and the box under it for paper and typewriter ribbons, and the little dispenser chute just under the space-bar, and acted like she was waiting for it to do something. But the next evening I went out to do some shopping, and I heard Susan playing with the keys when I came back.

Susan learned to type quickly, and even I learned enough to be able to send in the weekly reports— how much the food had cost, whether I had gotten married over the week, whether anyone was sick or the rent went up, and all the other things Mr. Stevens had asked us about. It took me about an hour at first to send a report, but after a while it became easier for me, and I spent a few hours each day practicing, with the switch off.

One year later your men came around again with paper and ribbons and a small heavy box. They told me they were adding something to the terminal, making it a teaching machine so I could learn better English and perhaps get a job someplace, and so my children could do better in school. They took a grey box with a lot of metal cards out of the box they had brought, and slipped it into the back of the machine, then asked me to type something. I wrote the sentence, "Kathy & Susan's in school," and then the machine made a little whirring noise and typed out on the same paper I had been using, "KATHY AND SUSAN *ARE* AT SCHOOL." The men smiled and told me it was working fine, and that I should try to write the way the machine corrected me, because it had been set up by a very smart English teacher, and that later on they would attach things so we could ask it questions in history or mathematics or even the news and it would help us to learn those things.

And that sure was the truth— my girls got much better grades with that machine drilling them all the time, and I learned how to write just as well as that typewriter, in only another year, but I didn't get a job because I didn't learn anything else. But I thought about it, and one day I turned the switch on and typed into the machine asking it whether it thought I should get a job or should stay home and take care of Kathy and Susan— and with handling all our reports for so long, I guess the computer had gotten to know us very well, because it answered right back, "TAKE CARE OF KATHY AND SUSAN— BUT WELFARE DEPT HAS REQUISITION FOR SKILLS PROGRAMMING." That meant they were sending over some new boxes to teach me a skill, I found out, and in a few more months I was answering all the questions on being a saleslady just right. In a little while, I got a part-time job selling dresses just a few blocks away, which helped pay some of the bills. When I typed that into the machine, it actually congratulated me and printed out a monthly check that was only twenty dollars smaller than normal!

Well, you can see that that terminal and I were getting on very good terms with each other. It got to a point where I didn't even have to think about it being a machine and not a person, because it always answered in a very understanding tone— I began to think that if it had been speaking when it was typing back to me, it might have sounded a lot like Mr. Stevens— or even a little like my husband. I suppose that had all been built into it, but just the same when I was alone sometimes, it was almost conforting to be able to talk with a terminal, tell it what had happened to me that day, and even ask how it was getting along. And its answers were always very nicely put.

Sometimes, when I'd been full of self-pity, and told the terminal how sad I was that my husband had left me, or even that I had been born in the first place, it would answer back in the same tones it had used when it was educating us, very stern, and tell me to cut out that nonsense— or it would just type out a correct answer, like, "I AM GLAD I WAS BORN— I HAVE KATHY AND SUSAN TO TAKE CARE OF, AND THEY ARE BOTH SO WONDERFUL— EVEN THE SKY IS ESPECIALLY PRETTY TODAY." That would usually make me feel a lot better— even when there was trouble at work, up to the day I was fired, I knew I could type it all in and there would be a comforting word for me, or sound guidance.

I was fired for falling asleep at work when I was the only one in the store. Some shoplifters came in and took about twenty dresses, and when I woke up, there was the manager and I couldn't explain a thing, or where the dresses went to, but I had to admit that I had been asleep, so they kept me on for two more weeks, and deducted the price of the dresses from both those checks, so that I didn't get anything, and then they fired me. I felt so bad that I decided never to get another job in this dirty world, and when I told that to the computer, I was so sad and so angry at my manager that the unemployment-sized check was printed right out, and then after it came another, for a hundred dollars! Then it sort of whirred and shut itself off.

I took the money and bought both of the children clothes, made a down payment on a small icebox, and even got some yellow material to fix the furniture. Susan's dress was especially pretty— it was a real formal, for her junior high school dance. She had been asked out by one of the boys on the baseball team, and she was so happy when we got her dress that she cried.

He was a senior, and he picked her up in his car. They drove to the dance, Susan waving goodbye to me and swinging her bright, soft hair like a little girl, and the boy, Robert, shaking my hand quite manfully and promising to be home early, then turning and taking Susan down the hallway to the staircase and then out into the street, opening the door for her and waiting until she was quite comfortable before he closed it, and then driving away from the curb so very cautiously, so very safely, that I knew she was safe with him.

At one in the morning I received a call from the hospital that Susan had been found in a car wreck with a boy and that they had both been rushed to the hospital, but would I come sign a release for her. They had been hit, according to the police, by a truck out of control.

That was just four days ago. I went to the hospital and signed a release, but wasn't allowed to see her or the boy, and was told to go back home. When I got back, Kathy was still asleep and I felt so tired and empty that I couldn't even cry— I just kept looking at Kathy and then I went over to that machine of yours and started to type as fast as I could, telling it everything about the accident and about Susan and Robert, and then everything about myself, and my childhood and then about my husband, and then I told it over again, and I was typing so fast I couldn't even tell that the machine was answering back on the same paper, with the same keys so that they were getting all tangled up, and then suddenly there came out this spray of printed checks from the chute underneath the space-bar into my lap, and the keys started clacking against each other on the paper, and the whole thing started to smell as though it were burning from trying to type at the same time I was and trying to print out checks, and the noise got louder and the machine started to smoke— this horrible oil smell filled up the room, and then there was a kind of "FZZT" and the terminal stopped dead.

Praying there was enough in the checks to cover Susan's hospital bills, I bent over to pick some of them up. My eyes were blurred and wet. I was exhausted; Susan was in the hospital and the terminal was broken. Kathy was awake and crying. But I bent down to pick up the checks and all there was was a great mess of type all over them, and not one of them good for a bank— half of them blank.

The hospital called me in the morning and told me that Susan was going to need an operation and a room in the hospital, and I told them we were on welfare but we'd try to get the money from them— you know we never could afford insurance— and to go ahead with it as quickly as they could. They did the operation two days ago, and I went to the hospital to see her, and she was unconscious with bandages all over her head and side, with her left arm broken and the room kept warm and dark. I cried all the time I was there, and a nurse had to drive me home.

This morning I got two letters— one from you people, asking me how the terminal broke down and enclosing this form to fill out. The other was from the hospital— a bill for three hundred dollars. Normally I would have told the computer all this, or even poor

Mr. Stevens, but if I am going to pay that bill, I have to get the money from you, so they don't throw Susan out of that hospital or try to move her to another— I've heard what happens to auto victims when they're moved.

So you see I must ask for the money in this letter, at the end of this form. I remember Mr. Stevens telling me that when things had to be done on paper they took a long time, that sometimes things got filed before they were even read, and even then the mail was so slow. I've tried waiting in your office but there are so many people there I waited all day without seeing anybody. Please, *please* give this requisition your fastest attention—

Yours urgently,
Mrs. Laura Martin

Ms Fnd in a Lbry
1961
Hal Draper

From: *Report of the Commander, Seventh Expeditionary Force, Andromedan Paleoanthropological Mission*

What puzzled our research teams was the suddenness of collapse, and the speed of reversion to barbarism, in this multi-galactic civilization of the biped race. Obvious causes like war, destruction, plague, or invasion were speedily eliminated. Now the outlines of the picture emerge, and the answer makes me apprehensive.

Part of the story is quite similar to ours, according to those who know our own prehistory well.

On the mother planet there are early traces of *books*. This word denotes paleoliterary records of knowledge in representational and macroscopic form. Of course, these disappeared very early, perhaps 175,000 of our yukals ago, when their increase threatened to leave no place on the planet's surface for anything else.

First they were reduced to *micros*, and then to *supermicros*, which were read with the primeval electronic microscopes then extant. But in another yukal the old problem was back, aggravated by colonization on most of the other planets of the local solar system, all of which were producing *books* in torrents. At about this time, too, their cumbersome alphabet was reduced to mainly consonantal elements (thus: *thr cmbrsm alfbt w rdsd t mnl cnsntl elmnts*) but this was done to facilitate quick reading, and only incidentally did it cut down the mass of *Bx* (the new spelling) by a full third. A drop out of the bucket.

Next step was the elimination of the multitude of separate Bx depositories in favor of a single building for the whole civilization. Every home on every inhabited planet had a farraginous diffuser which tuned in on any of the Bx at will. This cut the number to about one millionth at a stroke, and the wise men of the species congratulated themselves that the problem was solved.

This building, 25 miles square and two miles high, was buried in one of the oceans to save land surface for parking space, and so our etymological team is fairly sure that the archaic term liebury (lbry) dates from this period. Within no more than 22 yukals, story after story had been added till it extended a hundred miles into the stratosphere. At this level, cosmic radiation defarraginated the scanning diffusers, and it was realized that another limit had been reached. Proposals were made to extend the liebury laterally, but it was calculated that in three yukals of expansion so much of the ocean would be thus displaced that the level of the water would rise ten feet and flood the coastal cities. Another scheme was worked out to burrow deeper into the ocean bottom, until eventually the liebury would extend right through the planet like a skewer through a shashlik (a provincial Plutonian delicacy), but it was realized in time that this would be only a momentary palliative.

The fundamental advance, at least in principle, came when the representational records were abandoned altogether in favor of *punched supermicros*, in which the supermicroscopic elements were the punches themselves. This began the epoch of abstract recs— or Rx, to use the modern term.

The great breakthrough came when Mcglcdy finally invented mass-produced *punched molecules* (of any substance). The mass of Rx began shrinking instead of expanding. Then Gldbg proved what had already been suspected: knowledge was not infinite, and the civilization was asymptotically approaching its limits; the flood was leveling off. The Rx storage problem was hit another body-blow two generations later when Kwlsk used the Mcglcdy principle to develop the *notched electron*, made available for use by the new retinogravitic activators. In the ensuing ten yukals a series of triumphant developments wiped the problem out for good, it seemed:

(1) Getting below matter level, Shmt began by notching quanta (an obvious extension of Kwlsk's work) but found this clumsy. In a brilliant stroke he invented the *chipped quantum*, with an astronomical number of chips on each one. The Rx contracted to one building for the whole culture.

(2) Shmt's pupil Qjt, even before the master's death, found the chip unnecessary. Out of his work, ably supported by Drnt and Lccn, came the *nudged quanta*, popularly so called because a permanent record was impressed on each quantum by a simple vectorial pressure, occupying no subspace on the pseudosurface itself. A whole treatise could be nudged onto a couple of quanta, and

whole branches of knowledge could for the first time be put in a nutshell. The Rx dwindled to one room of one building.

(3) Finally— but this took another yukal and was technologically associated with the expansion of the civilization to intergalactic proportions— Fx and Sng found that quanta in hyperbolic tensor systems could be tensed into occupying the same spatial and temporal coordinates, if properly pizzicated. In no time at all, a quantic pizzicator was devised to compress the nudged quanta into overlapping spaces, most of these being arranged in the wide-open areas lying between the outer electrons and the nucleus of the atom, leaving the latter free for tables of contents, illustrations, graphs, etc.

All the Rx ever produced could now be packed away in a single drawer, with plenty of room for additions. A great celebration was held when the Rx drawer was ceremoniously installed, and glowing speeches pointed out that science had once more refuted pessimistic croakings of doom. Even so, two speakers could not refrain from mentioning certain misgivings ...

To understand the nature of these misgivings, we must now turn to a development which we have deliberately ignored so far for the sake of simplicity but which was in fact going on side by side with the shrinking of the Rx.

First, as we well know, the Rx in the new storage systems could be scanned only by activating the nudged or pizzicated quanta, etc. by means of a code number, arranged as an index to the Rx. Clearly the index itself had to be kept representational and macroscopic, else a code number would become necessary to activate *it*. Or so it was assumed.

Secondly, a process came into play of which even the ancients had had presentiments. According to a tradition recorded by Kchv among some oldsters in the remote Los Angeles swamps, the thing started when an antique sage produced one of the paleoliterary Bx entitled *An Index to Indexes* (or *Ix t Ix*), coded as a primitive I^2. By the time of the supermicros there were several Indexes to Indexes to Indexes (I^3), and work had already started on an I^4.

These were the innocent days before the problem became acute. Later, Index runs were collected in Files, and Files in Catalogs— so that, for example, $C^3F^5I^4$ meant that you wanted an Index to Indexes to Indexes to Indexes which was to be found in a certain File of Files of Files of Files of Files, which in turn was contained in a Catalog of Catalogs of Catalogs. Of course, actual numbers were much greater. This structure grew exponentially. The process of education consisted solely in learning how to tap the Rx for knowledge when needed. The position was well put indeed in a famous speech by Jzbl to the graduates of the Central Saturnian University, when he said that it was a source of great pride to him that although hardly anybody knew anything any longer, everybody now knew how to find out everything.

Another type of Index, the Bibliography, also flourished, side by side with the C-F-I series of the Ix. This B series was the province of the aristocracy of scholars who devoted themselves exclusively to Bibliographies of Bibliographies of ... well, at the point in history with which we are next concerned, the series had reached B^{437}. Furthermore, at every exponential level, some ambitious scholar branched off to a work on a History of the Bibliographies of that level. The compilation of the first History of Bibliography (H^1) is lost in the mists of time, but there is an early chronicled account of a History of Bibliographies of Bibliographies of Bibliographies (H^3) and naturally H^{436} was itself under way about the time B^{437} was completed.

On the other hand, the first History of Histories of Bibliographies came much later, and this H-prime series always lagged behind. It goes without saying that the B-H-H series (like the C-F-I series) had to have its own indexes, which in turn normally grew into a C-F-I series ancillary to the B-H-H series. There were some other but minor developments of the sort.

All these Index records were representational; though proposals were made at times to reduce the whole thing to pizzicated quanta, reluctance to take this fateful step long won out. So when the Rx had already shrunk to room-size, the Ix were expanding to fill far more than the space saved. The old liebury was bursting. One of the asteroids was converted into an annex, called the Asteroidal Storage Station. In thirteen yukals, all the ASS's were filled in the original solar system. Other systems selfishly refused to admit the camel's nose into their tent.

Under the stress of need, resistance to abstractionizing broke, and with the aid of the then new process of cospatial nudging, the entire mass of Ix was nudged into a drawer no bigger than that which contained the Rx themselves.

Now this drawer (D^1) had to be activated by indexed code numbers, itself. More and more scholars turned away from research in the thinner and thinner stream of discoverable knowledge in order to tackle the far more serious problem: how to thread one's way from the Ix to the Rx. This specialization led to a whole new branch of knowledge known as Ariadnology. Naturally, as Ariadnology expanded its Rx, its Ix welled proportionately, until it became necessary to set up a sub-branch to systematize access from the Ix to the Rx of Ariadnology itself. This (the Ariadnology of Ariadnology) was known as A^2, and by the time of the Collapse the field of A^5 was just beginning to develop, together with its appropriate Ix, plus the indispensable B-H-H series, of course.

The inevitable happened in the course of a few yukals: the Ix of the second code series began to accumulate in the same ASS's that had once been so joyfully emptied. Soon these Ix were duly abstractionized into a second drawer, D^2.

Then it was the old familiar story: the liebury filled up, the ASS's filled up. Around 10,000 yukals ago, the first artificial planet was created, therefore, to hold the steadily mounting agglomeration of Ix drawers. About 8000 yukals ago, a number of artificial planets were united into pseudosolar systems for convenience. By the time of yukal 2738 of our own era (for we are now getting into modern times), the artificial pseudosolar systems were due to be amalgamated into a pseudogalaxy of drawers, when— the Catastrophe struck ...

This tragic story can be told with some historical detail, thanks to the work of our research teams.

It began with what seemed a routine breakdown in one of the access lines from $D^{57 \times 103}$ to $D^{42 \times 107}$. A Bibliothecal Mechanic set out to fix it as usual. It did not fix. He realized that a classification error must have been made by the ariadnologist who had worked on the last pseudosolar system. Tracing the misnudged quanta involved, he ran into:

"See $C^{11}F^{73}I^{15}$."

Laboriously tracing through, he found the note:

"This Ix class has been replaced by $C^{32}F^7I^{10}$ for brachygravitic endo-ranganathans and $C^{22}F^{64}I^3$ for ailurophenolphthaleinic exoranganathans."

Tracing this through in turn, he found that they led back to the original $C^{11}F^{73}I^{15}$!

At this point he called in the district Bibliothecal Technician, who pointed out that the misnudged sequence could be restored only by reference to the original Rx. Through the area Bibliothecal Engineer, an emergency message was sent to the chief himself, Mlvl Dwy Smth.

Without hesitation, His Bibliothecal Excellency pressed the master button on his desk and queried the Ix System for: "Knowledge, Universal— All Rx-Drawers, *Location of.*"

To his stunned surprise, the answer came back: "See also $C^{11}F^{73}I^{15}$."

Frantically he turned dials, nudged quanta, etc. but it was no use. Somewhere in the galaxy-size flood of Ix drawers was the one and only drawer of Rx, the one that had once been installed with great joy. It was somewhere among the Indexes, Bibliographies, Bibliographies of Bibliographies, Histories of Bibliographies, Histories of Histories of Bibliographies, etc.

A desperate physical search was started, but it did not get very far, breaking down when it was found that no communication was possible in the first place without reference to the knowledge stored in the Rx. As the entire bibliothecal staff was diverted for the emergency, breakdowns in the access lines multiplied and tangled, until whole sectors were disabled, rendering further cooperation even less possible. The fabric of this biped civilization started falling apart.

The final result you know from my first report. Rehabilitation plans will be sent tomorrow.

Yours,
Yrlh Vvg
Commander

(Handwritten memo) This report received L-43-102. File it under $M^{42}A^8E^{39}$.— T.G.

(Handwritten memo) You must be mistaken; there is no $M^{42}A^8E^{39}$. Replaced by $*W-M^{23}A^{72}E^{30}$ for duodenomattoid reports.— L.N.

(Handwritten memo) You damfool, you bungled again. Now you've got to refer to the Rx to straighten out the line. Here's the correction number, stupid:

Moxon's Master
1893
Ambrose Bierce

"Are you serious?— do you really believe that a machine thinks?"

I got no immediate reply; Moxon was apparently intent upon the coals in the grate, touching them deftly here and there with the fire poker till they signified a sense of his attention by a brighter glow. For several weeks I had been observing in him a growing habit of delay in answering even the most trivial of commonplace questions. His air, however, was that of preoccupation rather than deliberation: one might have said that he had "something on his mind."

Presently he said:

"What is a 'machine'? The word has been variously defined. Here is one definition from a popular dictionary: 'Any instrument or organization by which power is applied and made effective, or a desired effect produced.' Well, then, is not a man a machine? And you will admit that he thinks— or thinks he thinks."

"If you do not wish to answer my question," I said, rather testily, "why not say so?— all that you say is mere evasion. You know well enough that when I say (machine) I do not mean a man, but something that man has made and controls."

"When it does not control him," he said, rising abruptly and looking out of a window, whence nothing was visible in the blackness of a stormy night. A moment later he turned about and with a smile said: "I beg your pardon; I had no thought of evasion. I considered the dictionary man's unconscious testimony suggestive and worth something in the discussion. I can give your question a direct answer easily enough; I do believe that a machine thinks about the work that it is doing."

That was direct enough, certainly. It was not altogether pleasing, for it tended to confirm a sad suspicion that Moxon's devotion to study and work in his machine-shop had not been good for him. I knew, for one thing, that he suffered from insomnia, and that is no light affliction. Had it affected his mind? His reply to my question seemed to me then evidence that it had; perhaps I should think differently about it now. I was younger then, and among the blessings that are not denied to youth is ignorance. Incited by that great stimulant to controversy, I said:

"And what, pray, does it think with— in the absence of a brain?"

The reply, coming with less than his customary delay, took his favorite form of counter-interrogation:

"With what does a plant think— in the absence of a brain?"

"Ah, plants also belong to the philosopher class! I should be pleased to know some of their conclusions; you may omit the premises."

"Perhaps," he replied, apparently unaffected by my foolish irony, "you may be able to infer their convictions from their acts. I will spare you the familiar examples of the sensitive mimosa, the several insectivorous flowers and those whose stamens bend down and shake their pollen upon the entering bee in order that he may fertilize their distant mates. But observe this. In an open spot in my garden I planted a climbing vine. When it was barely above the surface I set a stake into the soil a yard away. The vine at once made for it, but as it was about to reach it after several days I removed it a few feet. The vine at once altered its course, making an acute angle, and again made for the stake. This maneuver was repeated several times, but finally, as if discouraged, the vine abandoned the pursuit and ignoring further attempts to divert it traveled to a small tree, further away, which it climbed.

"Roots of the eucalyptus will prolong themselves incredibly in search of moisture. A well-known horticulturist relates that one entered an old drain pipe and followed it until it came to a break, where a section of the pipe had been removed to make way for a stone wall that had been built across its course. The root left the drain and followed the wall until it found an opening where a stone had fallen out. It crept through and following the other side of the wall back to the drain, entered the unexplored part and resumed its journey."

"And all this?"

"Can you miss the significance of it? It shows the consciousness of plants. It proves that they think."

"Even if it did— what then? We were speaking, not of plants, but of machines. They may be composed partly of wood— wood that has no longer vitality— or wholly of metal. Is thought an attribute also of the mineral kingdom?"

"How else do you explain the phenomena, for example, of crystallization?"

"I do not explain them."

"Because you cannot without affirming what you wish to deny, namely intelligent cooperation among the constituent elements of the crystals. When soldiers form lines, or hollow squares, you call it reason. When wild geese in flight take the form of a letter V you say instinct. When the homogeneous atoms of a mineral, moving freely in solution, arrange themselves into shapes mathematically perfect, or particles of frozen moisture into the symmetrical and beautiful forms of snowflakes, you have nothing to say. You have not even invented a name to conceal your heroic unreason."

Moxon was speaking with unusual animation and earnestness. As he paused I heard in an adjoining room known to me as his "machine-shop," which no one but himself was permitted to enter, a singular thumping sound, as of some one pounding upon a table with an open hand. Moxon heard it at the same moment and, visibly agitated, rose and hurriedly passed into the room whence it came. I thought it odd that any one else should be in there, and my interest in my friend— with doubtless a touch of unwarrantable curiosity— led me to listen intently, though, I am happy to say, not at the keyhole. There were confused sounds, as of a struggle or scuffle; the floor shook. I distinctly heard hard breathing and a hoarse whisper which said "Damn you!" Then all was silent, and presently Moxon reappeared and said, with a rather sorry smile:

"Pardon me for leaving you so abruptly. I have a machine in there that lost its temper and cut up rough."

Fixing my eyes steadily upon his left cheek, which was traversed by four parallel excoriations showing blood, I said:

"How would it do to trim its nails?"

I could have spared myself the jest; he gave it no attention, but seated himself in the chair that he had left and resumed the interrupted monologue as if nothing had occurred:

"Doubtless you do not hold with those (I need not name them to a man of your reading) who have taught that all matter is sentient, that every atom is a living, feeling, conscious being. *I* do. There is no such thing as dead, inert matter; it is all alive; all instinct with force, actual and potential; all sensitive to the same forces in its environment and susceptible to the contagion of higher and subtler ones residing in such superior organisms as it may be brought into relation with, as those of man when he is fashioning it into an instrument of his will. It absorbs something of his intelligence and purpose— more of them in proportion to the complexity of the resulting machine and that of its work.

"Do you happen to recall Herbert Spencer's definition of 'Life'? I read it thirty years ago. He may have altered it afterward, for anything I know, but in all that time I have been unable to think of a single word that could profitably be changed or added or removed. It seems to me not only the best definition, but the only possible one.

" 'Life,' he says, 'is a definite combination of heterogeneous changes, both simultaneous and successive, in correspondence with external coexistences and sequences.' "

"That defines the phenomenon," I said, "but gives no hint of its cause."

"That," he replied, "is all that any definition can do. As Mill points out, we know nothing of cause except as an antecedent— nothing of effect except as a consequent. Of certain phenomena, one never occurs without another, which is dissimilar: the first in point of time we call cause, the second, effect. One who had many times seen a rabbit pursued by a dog, and had never seen rabbits and dogs otherwise, would think the rabbit the cause of the dog.

"But I fear," he added, laughing naturally enough, "that my rabbit is leading me a long way from the track of my legitimate quarry: I'm indulging in the pleasure of the chase for its own sake. What I want you to observe is that in Herbert Spencer's definition of 'life' the activity of a machine is included— there is nothing in the definition that is not applicable to it. According to this sharpest of observers and deepest of thinkers, if a man during his period of activity is alive, so is a machine when in operation. As an inventor and constructor of machines I know that to be true."

Moxon was silent for a long time, gazing absently into the fire. It was growing late and I thought it time to be going, but somehow I did not like the notion of leaving him in that isolated house, all alone except for the presence of some person of whose nature my conjectures could go no further than that it was unfriendly, perhaps malign. Leaning toward him and looking earnestly into his eyes while making a motion with my hand through the door of his workshop, I said:

"Moxon, whom have you in there?"

Somewhat to my surprise he laughed lightly and answered without hesitation:

"Nobody; the incident that you have in mind was caused by my folly in leaving a machine in action with nothing to act upon, while I undertook the interminable task of enlightening your understanding. Do you happen to know that Consciousness is the creature of Rhythm?"

"O bother them both!" I replied, rising and laying hold of my overcoat. "I'm going to wish you good night; and I'll add the hope that the machine which you inadvertently left in action will have her gloves on the next time you think it needful to stop her."

Without waiting to observe the effect of my shot I left the house.

Rain was falling, and the darkness was intense. In the sky beyond the crest of a hill toward which I groped my way along precarious plank sidewalks and across miry, unpaved streets I could see the faint glow of the city's lights, but behind me nothing was visible but a single window of Moxon's house. It glowed with what seemed to me a mysterious and fateful meaning. I knew it was an uncurtained aperture in my friend's "machine-shop," and I had little doubt that he had resumed the studies interrupted by his duties as my instructor in mechanical consciousness and the fatherhood of Rhythm. Odd, and in some degree humorous, as his convictions seemed to me at that time, I could not wholly divest myself of the feeling that they had some tragic relation to his life and character— perhaps to his destiny— although I no longer entertained the notion that they were the vagaries of a disordered mind. Whatever might be thought of his views, his exposition of them was too logical for that. Over and over, his last words came back to me: "Consciousness is the creature of Rhythm." Bald and terse as the statement was, I now found it infintely alluring. At each recurrence it broadened in meaning and deepened in suggestion. Why, here, (I thought) is something upon which to found a philsopy. If consciousness is the product of rhythm all things *are* conscious, for all have motion, and all motion is rhythmic. I wondered if Moxon knew the significance and breadth of his thought— the scope of this momentous generalization; or had he arrived at his philosophic faith by the tortuous and uncertain road of observation?

That faith was then new to me, and all Moxon's expounding had failed to make me a convert; but now it seemed as if a great light shone about me, like that which fell upon Saul of Tarsus; and out there in the storm and darkness and solitude I experienced what Lewes calls "The endless variety and excitement of philosophic thought." I exulted in a new sense of knowledge, a new pride of reason. My feet seemed hardly to touch the earth; it was as if I were uplifted and borne through the air by invisible wings.

Yielding to an impulse to seek further light from him whom I now recognized as my master and guide, I had unconsciously turned about, and almost before I was aware of having done so found myself again at Moxon's door. I was drenched with rain, but felt no discomfort. Unable in my excitement to find the doorbell I instinctively tried the knob. It turned and, entering, I mounted the stairs to the room that I had so recently left. All was dark and silent; Moxon, as I had supposed, was in the adjoining room— the "machine-shop." Groping along the wall until I found the communicating door I knocked loudly several times, but got no response, which I attributed to the uproar outside, for the wind was blowing a gale and dashing the rain against the thin walls in sheets. The drumming upon the shingle roof spanning the unceiled room was loud and incessant.

I had never been invited into the machine-shop— had, indeed, been denied admittance, as had all others, with one exception, a skilled metal worker, of whom no one knew anything except that his name was Haley and his habit silence. But in my spiritual exaltation, discretion and civility were forgotten and I opened the door. What I saw took all philosophical speculation out of me in short order.

Moxon sat facing me at the farther side of a small table upon which a single candle made all the light that was in the room. Opposite him, his back toward me, sat another person. On the table between the two was a chessboard; the men were playing. I knew little of chess, but as only a few pieces were on the board it was obvious that the game was near its close. Moxon was intensely interested— not so much, it seemed to me, in the game as in his antagonist, upon whom he had fixed so intent a look that, standing though I did directly in the line of his vision, I was altogether unobserved. His face was ghastly white, and his eyes glittered like diamonds. Of his antagonist I had only a back view, but that was sufficient; I should not have cared to see his face.

He was apparently not more than five feet in height, with proportions suggesting those of a gorilla— a tremendous breadth of shoulders, thick, short neck and broad, squat head, which had a tangled growth of black hair and was topped with a crimson fez. A tunic of the same color, belted tightly to the waist, reached the seat— apparently a box— upon which he sat; his legs and feet were not seen. His left forearm appeared to rest in his lap; he moved his pieces with his right hand, which seemed disproportionately long.

I had shrunk back and now stood a little to one side of the doorway and in shadow. If Moxon had looked farther than the face of his opponent he could have observed nothing now, except that the door was open. Something forbade me either to enter or to retire, a feeling— I know not how it came— that I was in the presence of an imminent tragedy and might serve my friend by remaining. With a scarcely conscious rebellion against the indelicacy of the act I remained.

The play was rapid. Moxon hardly glanced at the board before making his moves, and to my unskilled eye seemed to move the piece most convenient to his hand, his motions in doing so being quick, nervous and lacking in precision. The response of his antagonist, while equally prompt in the inception, was made with a slow, uniform, mechanical and I thought, somewhat theatrical movement of the arm, that was a sore trial to my patience. There was something unearthly about it all, and I caught myself shuddering. But I was wet and cold.

Two or three times after moving a piece the stranger slightly inclined his head, and each time I observed that Moxon shifted his king. All at once the thought came to me that the man was dumb. And then that he was a machine— an automaton chessplayer! Then I remembered that Moxon had once spoken to me of having invented such a piece of mechanism, though I did not understand that it had actually been constructed. Was all his talk about the consciousness and intelligence of machines merely a prelude to eventual exhibition of this device— only a trick to intensify the effect of its mechanical action upon me in my ignorance of its secret?

A fine end, this, of all my intellectual transports— my "endless variety and excitement of philosophic thought!" I was about to retire in disgust when something occurred to hold my curiosity. I observed a shrug of the thing's great shoulders, as if it were irritated: and so natural was this— so entirely human— that in my new view of the matter it startled me. Nor was that all, for a moment later it struck the table sharply with its clenched hand. At that gesture Moxon seemed even more startled than I: he pushed his chair a little backward, as in alarm.

Presently Moxon, whose play it was, raised his hand high above the board, pounced upon one of his pieces like a sparrow-hawk and with the exclamation "checkmate!" rose quickly to his feet and stepped behind his chair. The automaton sat motionless.

The wind had now gone down, but I heard, at lessening intervals and progressively louder, the rumble and roll of thunder. In the pauses between I now became conscious of a low humming or buzzing which, like the thunder, grew momentarily louder and more distinct. It seemed to come from the body of the automaton, and was unmistakably a whirring of wheels. It gave me the impression of a disordered mechanism which had escaped the repressive and regulating action of some controlling part— an effect such as might be expected if a pawl should be jostled from the teeth of a ratchet-wheel. But before I had time for much conjecture as to its nature my attention was taken by the strange motions of the automaton itself. A slight but continuous convulsion appeared to have possession of it. In body and head it shook like a man with palsy or an ague chill, and the motion augmented every moment until the entire figure was in violent agitation. Suddenly it sprang to its feet and with a movement almost too quick for the eye to follow shot forward across table and chair, with both arms thrust forth to their full length— the posture and lunge of a diver. Moxon tried to throw himself backward out of reach, but he was too late: I saw the horrible thing's hands close upon his throat, his own clutch its wrists. Then the table was overturned, the candle thrown to the floor and extinguished, and all was black dark. But the noise of the struggle was dreadfully distinct, and most terrible of all were the raucous, squawking sounds made by the strangled man's efforts to breathe. Guided by the infernal hubbub, I sprang to the rescue of my friend, but had hardly taken a stride in the darkness when the whole room blazed with a blinding white light that burned into my brain and heart and memory a vivid picture of the combatants on the floor, Moxon underneath, his throat still in the clutch of those iron hands, his head forced backward, his eyes protruding, his mouth wide open and his tongue thrust out; and— horrible contrast!— upon the painted face of his assassin an expression of tranquil and profound thought, as in the solution of a problem in chess! This I observed, then all was blackness and silence.

Three days later I recovered consciousness in a hospital. As the memory of that tragic night slowly evolved in my ailing brain I recognized in my attendant Moxon's confidential workman, Haley. Responding to a look he approached, smiling.

"Tell me about it," I managed to say, faintly— "all about it."

"Certainly," he said; "you were carried unconscious from a burning house— Moxon's. Nobody knows how you came to be there. You may have to do a little explaining. The origin of the fire is a bit mysterious, too. My own notion is that the house was struck by lightning."

"And Moxon?"

"Buried yesterday— what was left of him."

Apparently this reticent person could unfold himself on occasion. When imparting shocking intelligence to the sick he was affable enough. After some moments of the keenest mental suffering I ventured to ask another question:

"Who rescued me?"

"Well, if that interests you— I did."

"Thank you, Mr. Haley, and may God bless you for it. Did you rescue, also, that charming product of your skill, the automaton chess-player that murdered its inventor?"

The man was silent a long time, looking away from me. Presently he turned and gravely said:

"Do you know that?"

"I do," I replied; "I saw it done."

That was many years ago. If asked to-day I should answer less confidently.

Part 6.
Control of Behavior

The idea of modifying or controlling human behavior to secure the blessings of social stability is not new in fiction. In one form or another it is a major component of utopian and anti-utopian visions. Contemporary fiction has responded to advances in several areas of science and technology. Electrical engineering has given us devices for eavesdropping, spying, and monitoring physiological states at a distance. Expansion of the medical pharmacopoeia and refinement of surgical techniques have enhanced our ability to modify or induce specific modes of behavior. Writers like Huxley and Orwell were sensitive to the possibilities of using these developments in social engineering; but before the advent of computers the integrative mechanisms essential to centralized social control were not clearly defined. Computer-communications technology allows for drawing a more complete picture.

Effective, large-scale use of techniques for controlling human behavior depends on fast, accurate and reliable information-processing. The ability to observe, monitor, and to intervene therapeutically would be useless without the means for storing, ordering, transmitting, and retrieving information. What is more it must be possible to perform these information-processing functions rapidly enough to permit decisions to be made and actions to be taken before the observations cease to be valid. The genuine social organism requires real-time coordination of sensors and effectors. Computers coupled with high-speed data-transmission facilities have breathed new life into the concept of an organic society. Given the accomplishments of the past— effected with far more limited means— it is difficult to dismiss the troubled musings of contemporary writers. Long before the development of modern communications— radio, television, telegraph— spies and informers were used with moderate success to maintain control over subservient populations; and the dossier system of the relatively sedentary European monarchies supported a primitive but reasonably effective system of intelligence surveillance.

More recently we have seen what can be achieved with an integrated system of spies, dossiers, methods of persuasion, and instantaneous communications. But the computer promises more than a simple extension of global intelligence operations. Remote though the possibility may appear at the moment, the new technology could be used to manage and shape the human environment in all its intimate details. One need not imagine a seizure of the reigns of government by a power-mad cabal— in fact this is not the most probable course of events. As we acquiesce in the alienation of personal responsibility, we relinquish the power to mould our own culture. Professional managers relieve us of the burdens of citizenship. Medical practitioners dissipate our anxieties and ease our frustrations. Educators equip us for specialized roles as well-oiled cogs in an incomprehensible machine. And for the society's rejects there is the appropriate agency to care.

Direct control by computers is not essential to effective machine domination. Stories in which conscious computers rule in their own right are throwbacks to a less sophisticated age. Caidin's *The God Machine* has a computer practicing hypnosis. In Fairman's *I, The Machine* the computer acts directly on human consciousness. In *B.E.A.S.T.* Maine fantasies about demonic possession by a machine. If these tales seem unreal, it is less from their technological implausibility than it is from the suggestion of heroic action by both man and machine. Step by imperceptible step, machines— part human, part computer— insinuate themselves into the social infrastructure. Constraints on moral choice and action make heroism unthinkable.

The increasing use of computers in social services— education, health-care, welfare— provides the factual basis for many stories of machine control of human behavior. Automation in the service area has two aspects. Activities are coordinated by means of information-processing systems, and parts of the service functions themselves are performed by computers. Most large organizations use computers to process data for general administration. Employee and client records, and accounting operations are typically computerized. Schools, hospitals, law enforcement and welfare agencies as well as corporations depend on computers to accomplish these vital functions. Yet the computer is not confined to such subsidiary roles. Systems have been developed to aid or replace teachers, and automated diagnosis has achieved limited success in medical practice. What emerges in fiction is a marriage of the two aspects of automation— computers coordinate and dispense, so that both administrators and practitioners are replaced by machines.

One recurring nightmare consists of an elaborate extension of computer applications in health-care. Sladek's "The Happy Breed" is a fantasy of regression into a uterine existence. All human needs are attended to by "Therapeutic Environment Machines" which monitor behavior and act to induce desired physiological or psychic states. No one works; the business of life is happiness. In this world without pain, administered by a central computer complex called MEDCENTRAL, individuality is equivalent to neurosis. As the story unfolds, we see the last five "abnormals" gradually being cured and brought under total control by MEDCENTRAL. This story is not set in the very distant future, and thus serves to dramatize current practices in psychotherapy— the distinction between helping an individual to cope with problems and altering attitudes so that problems disappear is often dangerously blurred.

The power of mind-altering drugs, and techniques for psychological conditioning to control mass behavior appears to be tremendously amplified by the introduction of computer systems capable of coordinating large-scale social programs. The ominous possibilities of this technological arsenal are a major preoccupation of computer fiction. In Brunner's *The Jagged Orbit* computer-aided psychotherapy is an effective instrument of political control. Deviation from behavioral norms is treated as a psycho-medical problem. Alban's "cocoon society" in *Catharsis Central* is quite similar to Sladek's world without pain, except that Alban allows for a way out. The presence of an escape route — planned obsolescence, imperfections in program design, or institutional weaknesses— is characteristic of earlier, romantic literature. Heroic defiance seems to have given way to futile gestures. Dozois' "Machines of Loving Grace" is a case in point. Computerized behavioral monitoring and advanced medical techniques make suicide virtually impossible. Everyone is equipped with implanted monitoring devices which signal physiological changes to a central computer. When critical changes are observed, the solicitous machines go into action.

Direct intervention in the human organism is by no means the only form of manipulation evident in fiction. Methods for exploiting social and cultural institutions are also explored. Computers are used to create conditions for the controlled release of aggression, and automated religious exercises minister to mass insecurities. In Eklund's fantasy "Dear Aunt Annie," the memory of a catastrophic war leads to the formation of anti-violence clinics as a safeguard against the resurgence of man's dangerous, belligerent impulses. "Annie Enterprises," headed by a complex robot, resembles an advice-to-the-lovelorn operation, and serves as the focus of anti-violence therapy. As the story reveals, the wholesale suppression of basic human impulses may have debilitating side-effects. Emschwiller's "Hunting Machine" dwells on another facet of controlled violence. An automated tracking and hunting device facilitates sadistic practices. The hunt is an unreal form of violent play, and contributes to the brutalization of the hunters.

The seemingly miraculous feats of the computer make the passage from admiration to awe and finally to worship entirely believable. Science has acquired a quasi-religious status in some circles, so that the emergence of cults associated with its sacred artifacts is to be expected. Some fictional cults occur spontaneously; others are contrived to manipulate the ignorant or the powerless. The religious practices associated with the Beast of Chorowait in Scheckley's *Journay Beyond Tomorrow* exemplify the latter. Chorowait's founders feel their computerized marvel to be an essential, unifying force. By terrorizing the community, the Beast serves to insure a healthy respect for religion, since survival depends on joint action by priests and laymen. Of course, the Beast's academic creators make sure to program the creature not to harm professors. In Brunner's "Judas" robot-worship occurs as a relatively spontaneous phenomenon— the inevitable result of a machine dominated culture. Mankind becomes enslaved by its own technological creations. Sladek deals with yet another form of computer-based religion in *The Müller-Fokker Effect*. An android is designed to replace a popular evangelist to allow greater flexibility in scheduling. In this case, the ritual performances are part of a cynical business enterprise.

One of the more subtle forms of social control operates by fostering the illusion of participation in decision-making. Burdick's *The 480* presents a realistic account of electoral politics in the United States. Computer simulation of voting behavior enables the pundits to manipulate political events by pandering to different voting blocs at the same time. Armed with precise knowledge of popular prejudices, pseudo-issues can be created to induce the people of a given region to vote for a particular candidate. Real issues are never addressed, and the act of casting a ballot becomes an empty gesture. The practice of engineering individual choice in accordance with predetermined social goals is carried much further in the future society of Compton's *The Steel Crocodile*. Here the computer assists in anticipating the consequences of individual behavior so that measures can be taken to insure desired results long before anyone becomes aware that they are being manipulated.

The use of the computer as an instrument of social engineering on a large scale depends on the elaboration of computer-communications technology. It is no exaggeration to say that the computer network is one of the most critical technological developments in contemporary society. The presence in every home and business of remote terminals which are linked to a vast network of interconnected computers would offer an incalculable temptation to would-be manipulators. It is far from clear that our political system could in principle devise safeguards against potential abuses.

In his novel *Fistful of Digits* Hodder-Williams illuminates the insidious threat posed by computer-communications. On the surface Servex is a world-wide computer network under the control of a clandestine group using it to pursue unspecified selfish ends. In fact there is no effective control center— the network is a reflection of collective neuroses. The plug-in communications device used to condition Servex converts is a symbol of human dependence on machines. Exercise of human control is an illusion because Servex is an entity with an autonomous will. At issue in the struggle between Shackleton and Servex for possession of Christina is the survival of human freedom. What began presumably as just another tool subject to man's will evolved into an instrument of man's degradation.

Unfortunately, the problem of insuring against socially undesirable uses of computer networks does not reduce to the enactment of protective laws and regulations. Legislation cannot deal with a cultural predisposition to relinquish power and responsibility in exchange for the promise of a happier, carefree existence.

The Twenty-third Psalm– Modern Style
1961
Alan Simpson

The Lord is my external-internal integrative mechanism,
I shall not be deprived of gratification for my
 viscerogenic hungers or my need-dispositions,
He motivates me to orient myself towards a non-social
 object with affective significance,
He positions me in a non-decisional situation,
He maximizes my adjustment.

From
The 480
1964
Eugene Burdick

Curver sat at a desk in the unlabeled Thatch headquarters. Three of the college boys were busy operating the automatic typewriters, signing Curver's name to the letters that came off the roller and then stuffing them into envelopes.

Curver wondered what Kelly thought about New Hampshire. Kelly would hardly believe the emery boards and the calendars had done it. He and Dev had commissioned a Ph.D. candidate in Berkeley (keep things away from New York and Washington and prevent leaks) to simply show a number of cheap objects to a wide number of people and let them choose what they wanted. The emery board and the wallet-sized calendar card were the winners. But these gimmicks were really just to keep Kelly off balance.

"They weren't a loss, Mad," Dr. Devlin said. "The middle-class person will hold onto something that has value even if it carries a slogan he doesn't like. A few of them will even come to believe in the slogan. It's part of the acquisitive aspect of the American personality."

"Don't let's kid ourselves, Dev. It was the simulation that won for us. And it had better work in Wisconsin and California." If they could manage a big write-in in those two states, the timing would be perfect. They would have shown strength in a March primary, again in April and then the last one, California, just a month before the Convention. That way the country would see Thatch, without running a visible campaign, steadily building up steam.

"We've got to win big in California, Mad."

"I know, I know," Curver said. The California one had to be the climax. "The Simulations people have been trying to simulate California for months. It won't be as easy as New Hampshire."

Curver got out the worksheets on the New Hampshire. He smoothed out the computer reports and then the interpretations he had made of them. First, he had told them to find groups that could be persuaded to cast a write-in vote and, second, a write-in vote for someone other than Lodge. He had assumed that Lodge would get close to 10,000 with a moderately successful campaign and more if his people really poured in money, just because he looked like a favorite son.

The instructions to the computer were few and simple: Locate people who had moved into the state recently, people who had moved out and then back to the state, people with low political information, people who were hostile or at least resentful of the regular Republican Party.

The results had been interesting. The newcomers lived mostly in the urban fringe around Manchester and a few other cities and the urban-fringe population had increased by 104.1 per cent in the ten years between 1950 and 1960. The group with low political information and some evidence of past hostility to the regular Republican Party were the blue-collar workers and there was a higher percentage of them in New Hampshire than in almost any other state. There was a solid group of women over sixty-five who were working and a substantial percentage of these would probably have a residual dislike of the rich-people image of the Republican Party.

The next instruction card was more difficult to formulate. Mad had wanted to find groups who felt deeply about some particular issue. This meant feeding to the computer information which was not directly political ... how many people were unemployed, how many unwed mothers, how many people sought a better mental-health program, and so on. A good deal of the information, fortunately, was available right from census tapes, government records, and specialized polls.

As the answers started to come in, the strategy became clear. One segment of people disliked the fact that both Rockefeller and Goldwater were millionaires but were trying to talk folksy. The anti-Communist anxiety still ran high. The older people were both suspicious and envious of the frivolous life in the hot Western states. A streak of Calvinism ran strong through the people and was reinforced by a long tradition of driving hard deals with the summer tourists. Mad concluded they had approved the state lottery because they thought most of the tickets would be bought by visitors from other states.

"I'd hate to have to show that throwaway to Bookbinder. He would not only think it was crazy, but he'd wince at the crudity of the whole thing," Curver said.

Dr. Devlin was glancing at the throwaway.

"It doesn't matter what Bookbinder likes," she said. "I don't think he'd know a typical voter if he met one on the street."

"What if Kelly analyzes that letter and starts to add things up?"

"No," Dr. Devlin said. "Kelly's like Bookbinder. He never really gets out and sees a voter. He deals with the influential people in the Party. They like style in a speech or a letter. But style isn't what gets some blue-collar worker with a seventy-two-year-old mother who is working in a shoe factory to write in someone's name. Your letter had something in it for a number of groups we thought might be pried loose from their traditional voting habit.

"They just picked out the sentence or two that applied to them and let it go at that." She glanced at him almost admiringly. "And it worked. The pros know Lodge won't carry weight outside of New Hampshire and the man they are watching is Thatch."

"Unless Lodge wanted to come out and run and it's already too late for that," Curver said. "By June New Hampshire will be only a faint memory."

Machines of Loving Grace
1972
Gardner R. Dozois

Dawn was just beginning to color the sky. She huddled inside the small bathroom— door closed, bolt slid and locked— sitting on the toilet lid and hugging her knees. Her head was tilted and hung down, chin almost on breast, and her eyes were nearly closed. She had wrapped her hands around her ankles. Her fingers were turning white. There was no noise in the empty apartment, not even the scurry of a cockroach. She had stopped crying hours ago.

There was noise beyond the window on her left, beyond plaster and glass, outside the vacuum of bedroom-kitchen-livingroom-guestroom-bath: a frozen automobile horn had been honking steadily for the last hour, occasionally traffic whined on the asphalt below, earlier in the evening there had been radios in nearby buildings, tuned to the confusion of a dozen different stations and fading one by one toward morning. She didn't pay any attention to these noises. The silence inside her apartment was too loud.

She opened her hands, flexed her stiff fingers, let her legs uncurl. One of them had gone to sleep, and she stamped it softly, automatically, to restore circulation. The floor was cold under her bare feet. Gooseflesh blossomed along her arms and she ran her hands down over them to smooth it. She had put on a new half-slip for the occasion. She shifted her weight; the toilet lid had been chilly at first, but now it had grown hot and sticky with the heat of her body. She leaned in closer to the hot-water pipe that descended from ceiling to floor— it was still warm to the touch. The dull paint had flaked off it in jigsaw pieces. There was a dingy grey toilet brush leaning against the base of the pipe. The bristles were broken and matted down. All this without thinking at all.

To be free, she thought, to keep my sanity.

Her head came up; eyes snapped open, closed to slits, opened again, wider.

The muscles in her neck had started to cord.

Her head jerked to the left. She stared out the window. Dawn was a growing red wash across the horizon, clustered buildings blocky beast-silhouettes, a factory plume of smoke etched black against tones of scarlet. Lights far away and lonely. A television antenna like a cross of stark metal. Her head turned back to center, wobbling: the string cut.

For a while she did not think. The shaving mirror on the wall over the sink, clutter on the shelves to the right of the basin: empty bottles of mouthwash, witch hazel, deodorant, the cardboard center from a roll of toilet paper, crumpled toothpaste tube, box of vaginal suppositories. The burlap curtains, frayed edges polarizing in the new light. Cracked and chipped plaster around the edges of the windowsill, streaks of white on the walls where paint had run thin. The closed door, the whorls in dark wood: beyond were the cluttered kitchen, the empty bedroom. They pressed in against the door. The door hinges were made in five sections.

I'm going to go crazy, she thought.

She reached out and flicked off the light switch. It was bright enough now to see: a gritty, hard light; harsh, too much grain and contrast. She had begun to tremble. The noise of the horn in the background was a steady buzz through her teeth. She picked up the razor blade from the window ledge. The horn stopped abruptly. In the silence, she could hear pigeons fluttering and cooing on the adjacent roof.

She turned the razor blade over in her fingers. The blade was smooth and sharp. No nicks in it, like the ones she used to shave her legs. She'd saved this one special. Orange sunlight refracted along the honed edge of the blade.

The bathtub was only inches away on her right, its head to the toilet. Without getting up, she leaned over, turned on the hot-water tap. Let the water run. This early it was reluctant: the water sputtered, the pipes knocked. But after a while it began to run hot. A thin wisp of steam. She put her arm under the hot water and sliced her wrist, holding the razor between thumb and forefinger. Clumsily, she switched hands and sliced her other wrist. Then she dropped the blade. Her wrists stung dully, and she felt a spreading warmth and wetness. She lifted her arms away from the water. Blood, welling up in thick clots, running down her arms toward the elbows.

To be free, she thought.

She sat with her arms held over the tub, palms up. Already it was better; the pressure that had been trying to turn her into someone else was receding. She wouldn't go crazy this time. She tilted her arms up to help the flow. She noticed that the shower curtain had a pattern of yellow swans and fountains on it, that there was a quarter-full plastic bottle of shampoo and a bit of melted soap in the bath shelf. A big glob of blood had splattered against the porcelain bottom of the tub. The flowing water stretched it out elastically, tugged at it, swept it loose and swirled it down the drain.

Too slow. The Lysol had been faster.

She fumbled for the razor blade, dropped it, wiped her hand dry on the shower curtain, picked it up again. She tilted her head back, felt for the big vein in her throat, located it with a finger. Very carefully, she positioned the razor blade. Then she closed her eyes and hacked with all her strength.

The control light flittered on the Big Board: green dulled to amber, died to red, guttered out completely. A siren began to scream. The duty tech put down his magazine, winced at the metallic wailing and touched the arm of his chair. Pneumatics hissed, the chair moved up and then sideways along the scaffolding, ghosting past thousands of unwinking green eyes set in horizontal rows, rows stacked in fifty-by-fifty-foot banks, banks filling the walls of the hexagonal Monitoring Complex, each tiny light in the walls in the banks in the rows representing the state of the life-system of one person in this sector of the City.

The tech found the deader easily: one blank spot in a solid wall of green— like a missing tooth, like the empty eye socket of a skull. He read the code symbols from the plaque above the dead light, relayed them through his throat mike to the duty runner down on the floor. "Got that?" "Check." Below, in Dispatching, the runner would be feeding the code symbols into a records computer, getting the coordinates of the deader's address, sending a VHF pulse out to the activated monitor in the deader's body, the monitor replying with a pulse of its own so that the computer could check by triangulation that the deader was actually at his home address and then flash confirmation to the runner. The whole process took about a minute. Then the runner, fingers racing over a keyboard, would relay the coordinates to the sophisticated robot brain of the meat wagon, flick the activating switch, and the pickup squad would whoosh out over the private government monorail system that webbed the City's roofs.

The duty tech hung from the scaffolding, twenty feet above the floor, three feet away from the banked lights of the Big Board. He settled back against the black leather cushions of his chair, waiting for the official confirmation. The siren had been cut off. He was bored. He nudged at the blank light with the toe of his shoe. Idly, he began to read the code symbols again. Somehow they seemed familiar.

The runner's voice buzzed in his head. "Dispatched." "Confirmed," the tech replied automatically, then still tracing the symbols with his finger: "Christ, do you know who this is? The deader? It's her again. That crazy broad. Christ, this is the third time this month."

"Fuck her. She's nuts."

The tech looked at the dead light, shook his head. The chair eased back down into its rest position before the metal desk. He squirmed around to get comfortable, drank the dregs of his coffee, rested his feet on the rim of the desk and settled back. The whole thing had taken maybe eight, maybe ten minutes. Not bad. He reached out and found the article he'd been reading.

By the time they brought her back, he was deep in the magazine again.

They carried her in and put her into the machines. The machines kept her in stasis to retard decay while they synthesized blood from sample cells and pumped it into her, grew new skin and tissue from scrapings, repaired the veins in throat and wrists, grafted the skin over them and flash-healed them without a scar. It took about an hour and a half, all told. It wasn't a big job. It was said that the machines could rebuild life from a sample as small as fifty grams of flesh, although that took a few weeks— even resurrect personality/identity from the psychocybernetic records for a brain that had been completely destroyed, although that was trickier, and might take months. This was nothing. The machines spread open the flesh of her upper abdomen, deactivated the monitor that was surgically implanted in every citizen in accordance with the law, and primed it again so that it would go off when her life-functions fell below a certain level. The machines sewed her up again, the monitor ticking smoothly inside her. The machines toned up her muscles, flushed out an accumulated excess of body poisons, burned off a few pounds of unnecessary fat, revitalized the gloss of her hair, upped her ratio of adrenaline secretion slightly, repaired minor tissue damage. The machines restarted her heart, got her lungs functioning, regulated her circulatory and respiratory systems, then switched off the stasis field and spat her into consciousness.

She opened her eyes. Above, a metal ceiling, rivets, phosphorescent lights. Behind, a mountain of smoothly chased machinery, herself resting on an iron tongue that had been thrust out of the machine: a rejected wafer. Ahead, a plastic window, and someone looking through it. Physically, she felt fine. Not even a headache.

The man in the window stared at her disapprovingly, then beckoned. Dully, she got up and followed him out. She found that someone had dressed her in street clothes, mismatched, colors clashing, hastily snatched from her closet. She had on two different kinds of shoes. She didn't care.

Mechanically, she followed him down a long corridor to a plush, overstuffed office. He opened the door for her, shook his head primly as she passed, closed it again. The older man inside the office told her to sit down. She sat down. He had white hair (bleached), and sat behind a huge mahogany desk (plastic). He gave her a long lecture, gently, fatherly, sorrowfully, trying to keep the perplexity out of his voice, the hint of fear. He said that he was concerned for her. He told her that she was a very lucky girl, even if she didn't realize it. He told her about the millions of people in the world who still weren't as lucky as she was. "Mankind is free of the fear of death for the first time in the history of the race," he told her earnestly, "at least in the Western world. Free of the threat of extinction." She listened impassively. The office was stuffy; flies battered against the closed windowpane. He asked her if she understood. She said that she understood. Her voice was dull. He stared at her, sighed, shook his head. He told her that she could go. He had begun to play nervously with a paperweight.

She stood up, moved to the door. "Remember, young lady," he called after her, "you're free now."

She went out quickly, hurried along a corridor, past a robot receptionist, found the outside door. She wrenched it open and stumbled outside.

Outside, she closed the door and leaned against it wearily. It was full daylight now. In between dirty banks of clouds, the sun beat pitilessly down on concrete, heat rising in waves, no shadows. The air was thick with smoke, with human sweat. It smelled bad, and the sharper reek of gasoline and exhaust bit into her nostrils. The streets were choked, the sidewalks thick with sluggishly moving crowds of pedestrians, jammed in shoulder to shoulder. The grey sky pressed down on her like a hand.

From
Journey Beyond Tomorrow
1962
Robert Sheckley

The Necessity For the Beast of the Utopia

In the morning, Joenes met with his colleagues from the University. He told them his adventures of the previous night and expressed indignation at not having been warned about the Beast.

"But my dear Joenes!" said Professor Hanley. "We wanted you to witness this vital facet of Chorowait for yourself, and to judge it without preconceptions."

"Even if that witnessing had cost me my life?" Joenes asked angrily.

"You were never in the slightest danger," Professor Chandler told him. "The Beast never attacks anyone connected with the University."

"It certainly seemed as though it was trying to kill me," Joenes said.

"I'm sure it *seemed* that way," Manisfree said. "But actually it was merely trying to get at Laka who, being a Chorowaitian, is a suitable victim for the Beast. You might have been jostled a bit when the Beast tore the girl from you."

Joenes felt chagrined at finding that his danger, which had seemed so dire the night before was now revealed as no danger at all. To conceal his annoyance, he asked, "What sort of creature was it and to what species does it belong?"

Geoffrard of Classics cleared his throat importantly and said, "The Beast you saw last night is unique, and should not be confused with the Questing Beast whom Sir Pellinore pursued, nor with the Beasts of Revelation. The Chorowaitian Beast is more closely akin to the Opinicus, which the ancients tell us was part camel, part dragon, and part lion, though we do not know in what proportions. But even this kinship is superficial. As I said, our Beast is unique."

Joenes asked, "Where did this Beast come from?"

The professors looked at each other and giggled like embarrassed schoolboys. Then Blake of Physics controlled his mirth and said to Joenes, "The fact of the matter is, we ourselves gave birth to the Beast. We constructed it part by part and member by member, using the Chemistry Lab on weekends and evenings. All departments of the University cooperated in the design and fabrication of the Beast, but I should especially single out the contributions made by Chemistry, Physics, Mathematics, Cybernetics, Medicine, and Psychology. And I must also mention the contributions of Anthropology and Classics, whose inspiration this was. Special thanks are due Professor Elling of Practical Arts who upholstered the entire Beast with the most durable of plastic skins. Nor should I forget Miss Hua, our student assistant, without whose careful collation of our notes the whole venture might have foundered."

The professors beamed happily at Blake's speech. Joenes, who had unwrapped a mystery only to find an enigma, still understood nothing.

Joenes said, "Let me see if I follow you. You *made* the Beast, constructing it out of ideas and inert matter in the Chemistry Lab?"

"That's very nicely put," Manisfree said. "Yes, that's exactly what we did."

"Was the Beast made with the knowledge of the University administration?"

Dalton winked and said, "You know how it is with those fellows, Joenes. They have an innate distaste for anything new, unless it's a gymnasium. So of course we didn't tell them."

"But they knew all the same," Manisfree said. "Administration always knows what's going on. But unless something is forced on their attention, they prefer to look the other way. They reason that a project like this might turn out well, in which case they and the University would get credit for farsighted wisdom. And if it turns out badly, they're safe because they knew nothing about it."

Several of the professors leaned forward with jokes about administrators on their lips. But Joenes spoke first, saying, "The construction of the Beast must have been very difficult."

"Indeed it was," said Ptolemy of Mathematics. "Excluding our own time, and the wear and tear on the Chem Lab, we had to spend twelve million four hundred thousand and twelve dollars and sixty-three cents on the fabrication of special parts. Hoggshead of Accounting kept a careful record of all expenses in case we should ever be asked."

"Where did the money come from?" Joenes asked.

"The government, of course," said Harris of Political Science. "I, and my colleague Finfitter of Economics, took over the problem of funds appropriation. We had enough left over to throw a victory banquet when Project Beast was completed. Too bad you weren't here for that, Joenes."

Harris forestalled Joenes's next question by adding, "Of course, we did not tell the government that we were building the Beast. Although they might still have granted funds, the inevitable bureaucratic delay would have been maddening. Instead, we said that we were working on a crash project to determine the feasibility of building an eight-lane coast-to-coast underground highway in the interests of national defense. Perhaps I do not need to add that Congress, which has always favored highway construction, voted immediately and enthusiastically to give us funds."

Blake said, "Many of us felt that such a highway would be eminently practical, and perhaps extremely necessary. The more we thought about it, the more the idea grew on us. But the Beast came first. And even with government funds at our disposal, the task was tremendously difficult."

"Do you remember," asked Ptolemy, "the excruciating problems of programming the Beast's computer brain?"

"Lord, yes!" Manisfree chuckled. "And what about the difficulties of giving it a parthenogenetic reproductive system?"

"Almost had us stopped," said Dalton. "But then, consider how we worked to coordinate and stablize the Beast's movements! The poor thing lurched around the lab for weeks before we got that right."

"It killed old Duglaston of Neurology," Ptolemy said sadly.

"Accidents will happen," Dalton said. "I'm glad we were able to tell Administration that Duglaston had gone on his sabbatical."

The professors seemed to have a thousand anecdotes about the building of the Beast. But Joenes impatiently broke into their reminiscences.

"What I wanted to know," Joenes said, "is *why* you built the Beast?"

The professors had to think for a moment. They were separated by many years from the ecstatic days when they had first discovered the reasons for the Beast. But luckily, the reasons were all still there. After a slight pause, Blake said:

"The Beast was necessary, Joenes. It or something exactly like it was needed for the success of Utopian Chorowait, and by extension, for the fulfillment of the future which Chorowait represents."

"I see," Joenes said. "But why?"

"It's really terribly simple," Blake said. "Consider a society like Chorowait, or any other society, and ask yourself what causes its dissolution. It's a difficult question, and there really is no answer. But we can't be content with that. Men *do* live in societies; it seems to be in their nature. Given that as a necessary condition, we wanted to build an ideal societal model at Chorowait. Since all societies are breaking down today, we wanted ours to be stable, and as equitable as possible within a framework of accepted democratic law. We also wanted a pleasant society, and a meaningful one. Do you agree that these are worthwhile ideals?"

"Certainly," Joenes said. "But the Beast— "

"Yes, here is where the Beast comes in. The Beast, you see, is the implicit necessity upon which Chorowait rests."

Joenes looked confused, so Blake went on:

"It's actually a simple matter, and can be understood very readily. But first you must accept the need for stability, equitability within a framework of accepted law, and a meaning for existence. This you have accepted. Next you must accept the fact that no society can be made to operate on mere abstractions. When virtue goes unrewarded and vice is unpunished, men cease to believe, and their society falls apart. I'll grant you that men need ideals; but they cannot sustain them in the valueless void of the present world. With horror men discover how very far away the gods are, and how little difference anything makes."

"We will also grant you," Manisfree said, "that the fault undoubtedly lies in the individual man himself. Even though he is a thinking being, he refuses to think. Though possessed of intelligence, he rarely employs it for his own betterment. Yes, Joenes, I think we can accept all that."

Joenes nodded, amazed at these points the professors had granted him.

"So, given all that," Blake said, "we now see the absolute necessity of the Beast."

Blake turned away as though everything had been said. But Dalton, more zealous, continued:

"The Beast, my dear Joenes, is nothing less than Necessity personified. Today, with all mountains climbed and all oceans plumbed, with the planets within reach and the stars much too far away, with the gods gone and the state dissolving, what is there left? Man must pit his strength against something; we have provided the Beast for him. No longer must man dwell alone; the Beast is forever lurking nearby. No longer can man turn against himself in his idleness; he must be forever alert against the depredations of the Beast."

Manisfree said, "The Beast makes Chorowait society stable and cohesive. If the people did not work together, the Beast would kill them one by one. Only by the efforts of the entire populace of Chorowait is the Beast kept in reasonable check."

"It gives them a healthy respect for religion," Dalton said. "One needs religion when the Beast is on the prowl."

"It destroys complacency," Blake said. "No one could be complacent in the face of the Beast."

"Because of the Beast," Manisfree said, "the community of Chorowait is happy, family-oriented, religious, close to the soil, and continually aware of the necessity for virtue."

Joenes asked, "What stops the Beast from simply destroying the entire community?"

"Programming," Dalton said.

"I beg your pardon?"

"The Beast has been programmed, which is to say, certain information and responses have been built into its artificial brain. Needless to add, we took a great deal of care over that."

"You taught the Beast not to kill University professors," Joenes said.

"Well, yes," Dalton answered. "We aren't too proud of that, to tell you the truth. But we thought we might be necessary for a while."

"How else is the Beast programmed?" Joenes asked.

"It is taught to seek out and destroy any ruler or ruling group of Chorowait people; next in priority to destroy the unvirtuous, and next to destroy any Chorowaitian. Because of that, any ruler must protect both himself and his people from the Beast. That in itself is quite enough to keep him out of mischief. But the ruler must also cooperate with the priesthood, without whose aid he is helpless. This serves as a decisive check to his powers."

"How can the priesthood help him?" Joenes asked.

"You yourself saw the witch doctor in action," Hanley said. "He and his assistants use certain substances that are gathered for them by the entire population of Chorowait. These, in proper combination, will turn the Beast back, since it is programmed to recognize and respond to the proper combination."

"Why can't the ruler simply take the substances and their combination, turn the Beast back himself, and rule without a priesthood?" Joenes asked.

"We took great care to preserve the separation between church and state," Harris said. "There is no single combination, you see, that will serve for all times the Beast appears. Instead, a vast quantity of formulae must be calculated each day, using lunar and stellar cycles, and variables such as temperature, humidity, wind speed, and the like."

"These calculations must keep the priests very busy," Joenes said.

"Indeed they do," Hanley said. "So busy that they have very little time in which to interfere with the affairs of the state. As a final safeguard against the possibility of a rich, complacent, and overweening priesthood, we have programmed a recurring random factor into the Beast. Against this nothing suffices, and the Beast will kill the witch doctor and no other. In that way, the witch doctor runs the same danger as does the ruler."

"But under those circumstances," Joenes aid, "why would anyone want to be a priest or ruler?"

"Those are privileged positions," Manisfree said. "And as you saw, the humblest villager also runs the risk of death from the Beast. Since this is the case, men with ability will always accept the greater danger in order to exercise power, to fight against the Beast, and to enjoy greater privileges."

"You can see the interlocking nature of all this," Blake said. "Both the ruler and the witch doctor maintain their positions only through the support of the people. An unpopular ruler would have no men to help him against the Beast, and would quickly be killed. An unpopular witch doctor would not receive the vital substances he needs in order to check the Beast, which must be gathered by the efforts of the entire people. Thus, both the ruler and the witch doctor hold power by popular consent and approval, and the Beast thus insures a genuine democracy."

"There are some interesting sidelights on all this," said Hanley of Anthropology. "I believe this is the first time in recorded history that the full range of magical artifacts has been objectively necessary for existence. And it is probably the first time there has ever been a creature on Earth that partook so closely of the supernatural."

"In view of the dangers," Joenes said, "I don't see why any of your volunteers stay on Chorowait Mountain."

"They stay because the community is good and purposeful," Blake said, "and because they can fight against a palpable enemy instead of an unseen madman who works by perversity and kills through boredom."

"Some few of our volunteers had their doubts," Dalton said. "They weren't sure they could stick it out, even though we convinced them of the rightness of the thing. For the uncertain ones, Doctor Broign of Psychology was able to devise a simple operation on the frontal lobes of the brain. This operation didn't harm them in any way, and did not destroy intelligence and initiative like the terrible lobotomies of the past. Instead, it simply wiped out all knowledge of a world outside of Chorowait. With that accomplished, they had no other place to go."

"Was that ethical?" Joenes asked.

"They volunteered of their own free will," Hanley said. "And all we took from them was a little worthless knowledge."

"We didn't like to do it," Blake said. "But the pioneer stage of any society is often marked by unusual problems. Luckily, our pioneer stage is almost at an end."

"It ceases," Manisfree said, "when the Beast spawns."

The professors paused for a moment of reverent silence.

"You see," Ptolemy said, "we went to considerable difficulty to make the Beast parthenogenetic. Thus, self-fertilizing, its unkillable spawn will quickly spread to neighboring communities. The offspring will not be programmed to stay within the confines of Chorowait Mountain, as the original Beast is. Instead, each will seek out and terrorize a community of its own."

"But other people will be helpless against them," Joenes said.

"Not for long. They will go to neighboring Chorowait for advice, and will learn the formulae for controlling their own particular Beast. In this way the communities of the future will be born, and will spread over the face of the earth."

"Nor do we plan to leave it simply at that," Dalton said excitedly. "The Beast is all very well, but neither it nor its offspring are completely safe against man's destructive ingenuity. Therefore we have obtained more government grants, and we are building other creations."

"We will fill the skies with mechanical vampires!" Ptolemy said.

"Cleverly articulated zombies will walk the earth!" said Dalton.
"Fantastic monsters will swim in the seas!" said Manisfree.
"Mankind shall live among the fabulous creations it has always craved," Hanley said. "The griffin and the unicorn, the monoceros and the martikora, the hippogriff and the monster rat, all of these and many others will live. Superstition and fear will replace superficiality and boredom; and there will be courage, too, in facing the djin. There will be happiness when the unicorn lays his great head in a virgin's lap, and joy when the Little People reward a virtuous man with a bag of gold! The greedy man will be infallibly punished by the coreophagi, and the lustful must beware of meeting the incarnate Aphrodite Pandemos. Man will no longer be alone in the universe, but will live with creatures as marvelous as himself. And he will live in accordance with the only rules his nature will accept— the rules that come from a supernatural made manifest upon the Earth!"

Joenes looked at the professors, and their faces glowed with happiness. Seeing this, Joenes did not ask if the rest of the world, outside of Chorowait, wanted this reign of the fabulous, or if they should perhaps be consulted about it. Nor did Joenes state his own impression, that this reign of the fabulous would be nothing more than a quantity of man-made machines built to act like the products of men's imaginations; instead of being divine and infallible, the machines would be merely mortal and prone to error, absurdly destructive, extremely irritating, and bound to be destroyed as soon as men had contrived the machinery to do so.

But it was not entirely a regard for his colleagues' feelings that stopped Joenes from saying these and other things. He also feared that such dedicated men might kill him if he showed a real spirit of dissent. Therefore he kept silent, and on the long ride back to the University he brooded on the difficulties of man's existence.

When they reached the University, Joenes decided that he would leave the cloistered life as soon as he possibly could.

From
Fistful of Digits
1968
Christopher Hodder-Williams

Dr. Norman Williams kicked the medicine trolley into the middle of the treatment room and Ham and Peter collected the chairs.

Stranger, looking grim, was still on the phone. Near him was the box of tricks that had been ejected from the car when Stan hit the power line.

In an undertone, Williams asked Peter: "You're quite certain there are no mikes in this section?"

"I've checked and double-checked. The nearest one is in the hotel foyer."

Williams gestured towards it. "Let's be quite sure. Is the door beyond this one shut?"

"They're both shut."

"Good."

Stranger hung up, sat down by the trolley, which comprised the conference table. "That was Dr. Irene Pellings," he said.

Williams' sharp, birdlike movement of the head registered cursory assessment. He realized there was more trouble, saw the suddenly fatigued drooping of the tell-tale mustache— litmus to all moods. Stranger was looking older, tireder; and in that moment Williams felt a pang of doubt. For if Stranger seemed to show a measure of incipient defeat, it could only be for good reason. Stranger continued: "A petition, from her language students, has been submitted, demanding her removal."

"On what grounds?"

"Does it matter?— you can do it at the pressing of a switchkey; and those students are sitting ducks every time they enter the language laboratory."

Peter recalled his own bitter experience and murmured: "The way things are run."

"Yes, sir ... and very efficient it is, too. The University are being quite sensible about it— sending her on leave to another college for a month. But the same thing will happen there."

Ham: "They'll run her out of business? Why?"

"Because she wiped those rigged teaching tapes; and because of a television broadcast she did, not long ago."

Williams asked Stranger if he'd seen it.

"Yes, sir." He looked down at the box of tricks. He seldom showed any overt hate. He did now. "I remember how she ended the broadcast— her exact words, in fact. She said: 'Many people fear computers, because they seem to impersonate human beings. But they are wrong. What they should fear is the opposite: human beings who impersonate computers.' It fits, of course."

Williams nodded, looked at his watch. "Well, it's late ... and I do want to raise this plan of Peter's. But we'll go round the table. Stranger has been doing some theoretical work and this would seem to relate to Pellings' suspension at the University. Richard ... ?"

Stranger squinted worriedly through tobacco smoke and spoke fast. "If Irene is right she's inadvertently provided the Text. Unfortunately, I don't have a worthy sermon. But for openers, here's a question: Has anybody thought what Christina is for ... ? Peter, what do you think she's for?"

"Must I reduce her to that level?"

Williams said impatiently: "Come on, this is too vital a moment for sentiment."

Peter picked up a pair of forceps and tapped them slowly on the trolley-top. Ham gazed down at the castors, as if they bore relevance to the matter in hand. The trolley wouldn't keep still; it was interesting to compute which way it would creep, according to who was leaning on it hardest. Ham thought, probably Peter was at that moment. "From the start," said Peter, "it was evidently vital for Servex to separate Christina from me. Ever since that test, way back in the States, they knew I was alien to them. No doubt they've kept track of me ever since— in fact, it's even possible that when John Forbes formed the company with me it was a deliberate move to maintain a close watch."

Ham said: "I go along with that assessment, but why did they need her so urgently?"

"Frankly I can't explain it. Perhaps someone else has an idea which might."

At such moments as these it was always to Stranger they looked. He never knew whether it was mathematics or sheer intuition that enabled him so often to rise to an occasion. But now, with the iron lung monotonously lifelining a patient in the cubicle with deep, measured phrases of breath, Stranger squinted beneath contorted brows at each pair of eyes in turn, then snapped: "Sirs. I think I know the answers—and you, Peter, won't like them. Sorry."

"Try me."

"You sure you want to hear it?"

"John ... ? We're not picking up a thing ... That's right; they must be in the hospital section. No mikes in there."

"You sure you want to hear it?"

Peter grinned. "That means you don't think we're going to believe you!"

Stranger didn't smile, concentrated utterly. "We have to think of a rapidly expanding computer system. We have to bear in mind the extent of the operations those computers are expected to carry out. Look at them! Factories all over the world; power dams and traffic control and air booking and nuclear power stations— and these are just the ones we actually know about. What about the ones we don't?— all the military and political and economic systems that we have every reason to believe are part of the vast Servex network ... and the way they link people, or get rid of people— like the passengers on the airliner, like the victims of the Trek, like Irene Pellings at the University. How do they build machines fast enough?"

Ham nodded. "I think I see where this is leading."

"Fellowes! Get weaving and dial up an Asking Channel to find out what's going on with the dinosaurs."

"Rawlins says we need authority for that."

"Then buzz him, you fool"

"Where I think it's leading, Ham, is to people like Christina."

Ham said impatiently: "How the hell can computers use her brain ... ? That's what you're suggesting, isn't it?— that having run out of computer capacity they're incorporating people-thinkers?" He grinned. "People-thinkers don't think very fast. Me, for instance!"

Stranger flicked ash like a maestro cueing an unresponsive section of the orchestra. "My dear Ham, 'people-thinkers' are part of every computer there is, in any case. If I do one operation on one machine, then cross the room and feed the result to another machine, I am a link between the two. My thought is: 'Take output from machine number one, and if appropriate, feed this to the input of machine number two'— an operation which requires human judgment."

Williams said tersely: "Ham is right. 'People-thinkers'— and don't blame me for that phrase!— can't act fast enough compared with the speed of the machinery. In the time it takes them to cross a room they have already become redundant."

Stranger wasn't thrown. "Exactly. But there's a way around that. You get the machines accustomed to whichever people-thinker you're using, and then you do something which provides you with all the human judgment, but working at a speed comparable with the rest of the machinery."

Peter looked startled. "What do you mean, 'the rest of the machinery'— ?"

Stranger chucked it straight at him. "You make a copy, of course!"

"A copy?" Peter half-rose from his chair. "You're not talking about Christina?"

"Listen, a minute ... ! I told you you weren't going to like it. But ask yourself why they need Christina so much! Couldn't it be because once they do all the intricate programming to get the computers adjusted to her particular personality-group, they can't just switch to someone else? Couldn't it be also that they can't have you interfering with her emotionally, simply because they know your personality-group is totally wrong for the machinery?"

"But! ... Look, I know I'm just gaping like an idiot, Richard; but how do you copy a human being? Have little tin ladies running around with clockwork motors?"

Stranger stared straight out of the window— no expression. "You build a simulator. That's right: you get hold of a volunteer who happens to have the right sort of basic brain ... Then you put this brain under a microscope. To find a suitable brain, you have some sort of place for screening the candidates— "

Ham suddenly snapped his fingers. "— Musiconics!"

Stranger took the point quietly. "Probably ... Then, as I say, you get this brain and you map it out and profile it and process it, and you do likewise to the machines so that they have the maximum facility for communication. Those candidates who won't do ... Well, we saw what happened to Stan." He turned to Peter. "Of course, you are a terrible nuisance to them. You love the prototype."

Peter yelled: "Prototype ... ! For Christ's sake, she's a human being!"

Ham, because he just could not face the issue, preferred to play mildly amused. Taking his glasses off to polish the lenses he said: "You're having us on! A prototype? What for? A production model?" He added: "Can I have one too, please?"

It was Dr. Williams who seemed to get the point— and Fat Mrs. Jemison would have hated him with all her bulging indignation at that moment. He went flushed in the face and crashed a fist on the trolley so the instruments hopped an inch. "No! He's right. But he doesn't mean it in the literal sense. Look here, think of what Dr. Pellings said! She said what we ought to fear are 'human beings who impersonate computers' ... in other words, that aspect of a human brain which is usable 'as' a computer. But you can't just plug in, like they do in the funnies." Another percussive crash on the trolley. "So why not make a copy?— if it works faster yet thinks the same way, they've got human judgment built into the machine!"

Ham just went right on with polishing the glasses. "Impossible. It would need to be about the size of the United Nations building."

"No. Modern solid-state circuitry is compact. Soon we'll be growing clumps of transistors in the garden."

Ham remained unmoved. "All right. Even supposing you could build such a monster ... How could you program it?"

This time, Stranger took over. "With Christina herself. How else?"

Ham did look up this time. "But that would imply mood: love, hate, fear, joy, not to mention retaliation."

"Sir! Are you suggesting Servex doesn't retaliate?"

"Fellowes? Look at this reading! It thinks the Freid must have found the mikes!"

"Yes. But It can't seem to make up Its mind."

"I know. We've simply got to get further with Christina's C.A.V. Is she ready–for the programming?"

"Rawlins told me this morning she needs another day's orientation."

"Bugger Rawlins!"

Williams to Stranger: "Do you think they used"— he indicated the box of tricks— "one of those things? I mean on her?"

Stranger handed it across. Williams plonked it on the trolley.

"These," said Williams, "hundreds of them— maybe even thousands— are being used on perfectly ordinary human beings who are becoming less ordinary every second as a result. We know roughly what they do. But we don't know where they're produced, or how they're distributed. Nor do we know who's using them. We've got to find out. Peter has a plan which may help us do just that."

Peter said: "We've got to find one of the Servex control depots. That's the only way of discovering how they operate."

Ham: "Can't we just trace the wires? I mean from the mikes they're using to bug us with?"

"No. They're using the power lines for that too. There's no way of tracing where the hell they lead. We've got to get Servex to react to something, so that something shows on the surface."

"Using those mikes?"

Williams said: "Precisely, Ham. And I want you to tell us the work you did on those reactions. For the game we're going to play is called 'Cat and Mouse'."

Ham nodded. "I get it ... Well, I don't know who's Tom and who's Jerry; but I did just happen to do some work back in California, with a professor who had views on 'the Electronic Revolution' ... " For a moment he seemed embarrassed at such pretensions to intellect. "Actually, I'm just a poor millionaire who inherited Big Daddy's airline. Weird things— if you can believe the papers— are happening to that airline; and even weirder things are happening to Daddy's millions— if you've read today's shares index... and that's excluding the alimony I pay out."

Politely, Williams interpolated: "Ham, do you think we could dispense with the slow jokes? I know you have useful things to say and you don't have to apologize ... What was this 'biological' theme you were elaborating on a few weeks ago, when you were too absorbed to remember to make fun of yourself?"

"Oh, that." Ham shifted in his chair, taking five seconds in which to erase the dry self-denigration. "Okay, I think it goes like this: You have to think of Servex as an organism ... a somewhat primitive organism whose nervous system has no brain center."

Stranger gestured extravagantly. "Primitive!" he cried. "That should please John Forbes no end."

"I meant," said Ham coldly, "the composite organism is primitive in the way it reacts."

Williams frowned. He never joked in conference ... and anyway the trolley kept shifting infuriatingly and was difficult to write on. So he heaved the Communications Unit on to the floor, which at least gave more space. "Continue."

"Okay ... the organism has certain lines of defense. For instance, it camouflages itself like the chameleon."

"Big business," said Peter.

"Sure. But there is also a more subtle disguise altogether. A back-to-front kind of a disguise. You see, by very virtue of calling itself something— the name Servex— it manages to convey the impression that it exists apart."

"Whereas," said Stranger, "it is really a part of Society."

"Yes ... a very big part. And unquestionably it intends to become the whole of Society. In other words, it represents a trend which is happening to Society itself. So we are not talking about a separate entity, but— in the final analysis— the whole of civilization. It's just that at this stage the thing hasn't caught on quite so well as, say, the mini-skirt." Ham pushed his glasses up over his brow, as if temporarily to erase this latter thought.

"Now, this organism ... Like many things in Society, it doesn't actually represent the interests of any one human being— in fact, it is in direct conflict with the interests of the individual. It is an imposed rule of thumb resulting from everyone knowing what everyone else is doing almost before they do it, and arises from the tightness of modern communications, which never allow anything to develop before everyone else is asked what they think. So ... the organism has no sense of direction. The tail wags the dog, as with tee-vee ratings that come in too fast for the program to establish itself.

"Then something stung it. We don't know what. But what happens when one animal is stung by another? Think of the squid. The squid sends a thick black oily substance into the water ... In other words, it hits back."

Ham, with his lethargic manner, didn't reveal the tension he felt beneath. But the trolley did. Somehow, Ham had got a pressure, one foot against the other, which involved a castor of the trolley. His foot slipped, and the trolley shot across the room, crashing finally against the drugs cupboard with a clatter of tin and shattering glass. Kidney basins and instruments alike shot into the air, then clattered on the floor, like the uproar of an accident staged in a comic radio show.

"Forbes? You down there?"

"What is it, Rawlins?"

"We have a prediction. It's not very clearcut, about eighty per cent reliable ... "

"Yes ... ?"

"You wanted to know whether they've found the mikes. The answer is that if they test us out, with some kind of plan of action— which they would discuss deliberately within range of a microphone— and if they do this within the next two hours, then they're on to us ... probably."

"What's that based on?"

"Mostly their change of behavior pattern. Their object, by the way, will be to get a reaction from you."

"You mean they want a meeting or something?"

"That is what it would mean, yes."

"Why all this uncertainty?"

"I don't know. But ever since the rainstorm the computers have detected some sort of bluff. It's causing a considerable overload and the degree of paradox is giving us a lot of trouble."

"But even if the Freid do this we'll still be only eighty per cent sure. That leaves twenty per cent doubt, doesn't it?"

"That's the best figure we can get and we're stuck with it."

"Well, Rawlins ... if they are trying to rig a meeting it suits us."

"John ... the computers don't like this game of Cat and Mouse."

"No matter! If the Freid want a meeting we could work on whoever they send. Get this man Peter Shackleton down here and wire him to a Communications Unit. Then send him back. In that way we could neutralize the Freid. Couldn't we?"

Williams picked up the surgical instruments and plopped them into the sterilizer. "Ham, did you do that on purpose?"

"You know darn well I didn't."

"Interesting. Unconsciously you express tension by applying certain forces to the trolley. The forces become unstable and the trolley goes out of control. There's evidence in the papers that this is what Servex is doing to its own internal electronics. If we pile on the tension this will accelerate."

"Yeah, but won't that be dangerous? Servex has a gigantic share of industry under its control. Won't we be making things worse if we deliberately set out to drive it crazy?"

"What's the alternative?" Wirily, Williams leaned on the windowsill, penetrating darkness with screwed-up eyes towards the hub of industrial Exeter. "All we'd be doing is to force people to bring their machinery under human control— and then leave it like that until they knew exactly what is causing what."

"It's a hell of a lot of responsibility to take on, and you could do an awful lot of damage if you miscalculated."

"So what do you do, Ham? Let it get worse and worse? We have to take a calculated risk. Leave it too late, and it would be like trying to hit the panic button long after the button could stop the machine. Think: once a natural process, like the flowing of a river, starts to push the machine which is supposed to be controlling the water-level, it doesn't matter what you do with that button; the disaster is assured."

"Hell, are you saying people have been sitting on their backsides ignoring red lights?"

"No. I'm saying they don't act on what they see happening to the river: if the little red light doesn't come on they remain hypnotized. They just sit there, drowning. In blind faith they believe the machine, they believe it is 'fail safe' because it has become their religion. There's only one way we can put a stop to that— "

"— and stop them we must!" said Stranger. "Force them back into using human control before there's no going back at all! What are you waiting for, Ham? Do you want the human race to be driven permanently out of its mind, by seeing the machines mimicking human behavior?— Some think that's happening even now! Or do we wait for an automated nuclear war, dreamed up by machinery that's equally demented?"

A silence ... till Ham, visibly affected, turned to Peter. "So what do we do?"

"What we do is to test them out: stage a 'radio play' within range of these microphones and pretend we're going to cut their trunk nerve!"

Stranger produced a blissful smile. "You mean, the Porthcurno cable ... ? What an exquisite idea!"

Part 7.
Human Vitality

Technology exerts a powerful influence on human thought and action. From the primitive tools of the stone age to the complex machine-based systems of the modern period, the instruments of *homo faber* have helped to shape culture as well as the material environment. Technology is the source for many of the concepts used to characterize human behavior. Everyday speech is filled with figures and similes comparing man with his tools; and important lines of philosophical and scientific inquiry have been sparked by technological metaphors. Since the seventeenth century the human being has been analyzed in terms of clockworks, heat engines, switching networks, servomechanisms, and information-processing systems. Questions suggested by these mechanistic models of organic functions have in some cases led to major scientific advances. The principles governing the operation of fuel-burning engines were successfully incorporated into models of energy transformations in living organisms. More recently information-processing concepts have been applied to the study of perception and cognition. Knowledge of physiological and psychological processes has grown, but the new insights are far from reassuring.

As the barriers to our understanding have fallen, so have the pillars of our world-view. Technology has shaken our belief in the uniqueness of human identity. We are no longer so confident that we alone were created in God's image— our own artifacts must share in the honors. Machines exist which surpass us in strength, speed, precision, reliability, and durability. What is more, intelligence is exhibited in the behavior of computer programs which prove mathematical theorems, play complicated games, recognize patterns, and perform other tasks normally thought to require human cognitive skills. The set of attributes defining man's unique endowment grows smaller and smaller, but the demonstrable accomplishments of technology are only partly responsible for the shrinkage. Our willingness to accept comparison on the machine's terms is equally important. When human behavior is modeled using concepts appropriate to machines, it is no wonder that similarities loom larger than differences. If for example we impose suitably restrictive conditions on the notions of thought, creativity, and consciousness, it is not unlikely that we will be able to design programs which realize these notions. One cannot be sure, however, that what has been discarded is not the better part of being human.

The literary response to machine encroachments on human identity reveals a struggle to reestablish a stable equilibrium— to reach an accommodation which recognizes the potentialities of men and machines as independent entities. This often takes the form of trying to vindicate the belief in man's special status. Machines are shown to be less versatile than humans, hampered by the lack of emotional experience, or flawed by their inability to err. In Blish's "Solar Plexus" a cyborg ship threatens to canabalize a man for needed spare parts. The computer's ignorance of the subtleties of interpersonal relationships leads it to reveal its own vulnerability, thereby giving a critical advantage to the intended victim. Boulle describes the creation of a series of intelligent machines in "The Perfect Robot." Professor Fontaine of the Electronic Brain Company builds a theorem prover, a chess player, and robots capable of sexual activity and reproduction. Ultimately he discovers the need to "unhinge" his intelligent creations in order to provide them with an artistic sense and a sense of humor. Balchin's "God and the Machine" focuses on the need for flexibility in human behavior. In this story a computer is programmed to play checkers, and wins consistently until its human opponent starts to play erratically. The computer is unable to cope with this situation since its program forbids both losing and cheating. Somehow the dilemma is resolved and the machine cheats. This behavior jars the computer's maker. He realizes that he was the victim of his own unyielding nature, and from then on becomes more human.

What these and other man-machine comparisons seem to suggest is the existence of an anomalous component in human intelligence— a kind of animal cunning— that is absent in machines. As a result humans are able to keep one step ahead of their computer adversaries. In Saberhagen's "Without a Thought" man faces "beserker," an ancient machine designed to destroy all living things. The beserker uses a "mind beam" to neutralize intelligence. Although this force works against humans, it has no effect on lower animals. The commander of the spaceship threatened by the beserker capitalizes on this property and manages to outwit his opponent. This view of human advantage sidesteps basic issues, for intelligent machines are capable of learning from experience and must inevitably acquire the ability to counteract strategems of human cunning.

The threat to human identity posed by machines does not always arise from overt conflict and confrontation. A much more disturbing aspect of this problem is the extent to which people take on machine characteristics. The anthropomorphizing impulse which imputes human attitudes to machines represents a kind of wish-fulfillment; the complementary tendency to think of ourselves in mechanistic terms reveals a type of narcissistic infatuation. We are seduced by a distorted image reflected in the machine, and moved by vanity to reformulate the nature of man. This is not an intellectual exercise— it is manifest in altered attitudes, values and expectations. In fiction, people are sometimes transformed into machines. Sheckley's "Can You Feel Anything When I Do This" highlights this transformation by means of a provocative reversal of roles. A robot Omnicleaner (Rom) on display in a department store falls in love with a lady shopper. Rom contrives to be shipped to the woman's home, and succeeds in arousing her by his attentions. The woman, however, turns out to be more of an automaton than the robot. Unable to express her feelings, she responds by denying them and wrenching out the robot's power cord. In Goulart's "Hardcastle" the computerized house is a symbol of the hostile solicitousness of machine technology. The computer seduces with the promise of comfort, convenience, and high style, but these offerings are not free. Compliance with a predetermined order takes precedence over freedom, individuality and self-respect; and the capacity to love dissolves in an artificial world of asexual mechanism.

Diminishing opportunities for significant choice and action lead inexorably to loss of ability. Human capabilities atrophy, vitality is sacrificed on the alter of security, and life becomes a hedonistic fantasy played out in a uterine void. Forster anticipated these possibilities in "The Machine Stops" long before computers appeared on the scene. In the sterile, air conditioned, underground civilization of the story, face-to-face interaction has been replaced by remote communications. Existence is sedentary and intellectualized, direct experience of the world is universally shunned, and all bodily needs are attended to by mechanical devices. When the machine fails, there is no one left who knows how to repair it, and the society perishes for want of the capacity to cope with the natural environment. The essentials of Forster's classic portrait of the future are reproduced in several computer tales. In Fairman's *I, the Machine* and Alban's *Catharsis Central*, similar degenerative changes result from dependence on a central computer.

Man's longing for immortality endows machines with an irresistible attraction. The relentless quest for the elixir of life animates our lust for mechanical perfection. To the children of the industrial revolution the deathlessness of machines is a galling paradox and an affront to human vanity. The narcissistic idolatry of technology springs from our furious desire to emulate machines and to reject the dominion of organic nature. We are no longer content to accept culture as the proper vehicle of immortality. Vanity bids us to exchange cultural continuity for the perpetual novelty of mechanical excitation. In "The Lost City of Mars" Bradbury takes us on a fantastic journey into the psyche. A wealthy eccentric gathers a heterogeneous group of people for an excursion through the newly refilled canals of Mars in search of a lost city. The party finds the fabled place, and in the several personal dramas which follow we witness the tragedy of man's vain attachments to machines. All but four perish in this automated nightmare of a city, victims of their own disordered fantasies. Part of the tragedy was that "The damn city does everything, which is too much!" Each of the survivors comes to a similar realization. The poet accepts the fact of his own mortality, and thereby affirms man's vital connection with nature.

The challenge of mechanistic culture sometimes reinforces a blind faith in the ultimate superiority of man. An indomitable, irrepressible human spirit redeems the glory of an heroic past. Most persistent is the romance of the rebel in utopia. Mechanisms for social control, formidable though they may be, are not perfect. Utopia is either powerless to eliminate or has its special reasons for tolerating small groups of disaffected individuals. In Levin's *This Perfect Day* there are quasi-atavistic settlements outside the Family. A small group of aliens maintain a precarious existence in the bowels of the monster-computer in Fairman's *I, the Machine*. Clarke's *The City and the Stars* allows for unpredictable change by having the Halls of Creation spew out "uniques" from time to time. Compton's futuristic society in *The Steel Crocodile* supports a class of "alienees." There are many variations on the theme, but a charismatic figure emerges and concerted action against the oppressive forces of automated society follows. In the rare cases of successful rebellion the victory is typically a Pyrrhic one. The destruction of a central computer, for example, leads to the disintegration of society. More often the machine regains the upper hand. In fact one sometimes has the uneasy feeling that the entire struggle is programmed by a very clever computer. Regardless of the outcome, some essential human quality is preserved, a sacred obligation has been discharged. Sisyphus is not crushed by the rock.

The claim to human superiority over machines is bound up with man's creative powers. Computer programs may be able to generate poetry, music, works of art, theorems, etc., but despite the articulation of technical excellence, the need for direct human expression is not satisfied. Ballard's "Studio 5, The Stars" is a reenactment of the myth of Melander and Corydon. In this modern version, the muse is offended by the computerization of poetry, and insists on a return to the true sources of inspiration. The poets are made to realize that they cannot trifle with creative genius, whereupon they abandon their machines. Their reaffirmation of the gift of poetic expression is an assertion of the inevitability of spiritual rebirth. Asimov's "The Feeling of Power" also addresses the question of creativity in an automated society. The rediscovery of arithmetic signals the beginning of liberation from the intellectual domination of the machine. Human thought cannot be forever confined by an alien mechanistic discipline.

The strongest testimonial to human vitality is contained in stories of the resurrection of man by machine. Herein lies a mystical belief in the cosmic purpose of the human race. Not only is man acknowledged the master by machines as in Aldiss'

"But Who Can Replace a Man" or in McIntosh's "Machine Made," but the human being is central to the unfolding of an ineluctable cosmic plan. In Zelazny's "For a Breath I Tarry" computers recreate man long after the human race has vanished from the earth. After roaming the dead planet for millennia, the computer-titans fulfill their destiny in the act of resurrection. Del Rey's "Instinct" explores the same compelling theme. The robots do not know precisely what it is they are trying to construct, nor are they fully aware of the object of their search. Some mysterious need drives them to experiment, and recognition of success is purely instinctual. Recreated man is the undisputed master. In Capek's 1921 play *R.U.R.* man's resurrection comes about in a different way. After destroying their human creators, the robots become human themselves. The myth of regeneration has a common conclusion— the exhortation of *R.U.R.*: "Go, Adam; go, Eve. The world is yours."

Hardcastle
1971
Ron Goulart

The house had a slight German accent.

Bob Lambrick had just landed his helicopter on the copter deck next to the low rambling ranch-style house and he was climbing down out of the ship, his portfolio and attaché case hugged under his left arm.

"I was about to kiss that orange tree goodbye," said the house from the speaker mounted in the bird feeder in one of the decorative pines beyond the landing area.

Bob glanced at the orange tree on his front quarter-acre. A lone orange was rolling across the bright grass and toward the edge of the hillside. It tumbled on over and fell two hundred feet down to the Pacific Ocean and Bob said, "I've done most of my flying in Westchester County. That's in New York State. I'm not used to California air currents yet, especially those between Carmel here and San Francisco."

"You really came close to that tree. I suppose they fly more flamboyantly back East. Particularly in New York. They're more liberal."

Bob nodded slowly in the direction of the tiny loudspeaker. He tapped the side of the copter with his free hand and silver flecks came off. "Scraped the paint a little, too. I came too close to that decorative grape arbor up on Camino Real. They shouldn't put grape arbors on top of highrise office buildings."

"You don't understand the California mystique yet, Mr. Lambrick," replied the house. "We're close to the earth out here, very nature-oriented. And, by the way, don't forget to wipe your feet."

Bob noticed the clods of mud on his commute boots. "I'll take them off and leave them out here." He set his briefcase and portfolio down and gave a tug at one of the boots.

"Stick your feet in the bootjack," suggested the house.

"Where is it?"

"Big cocoa-colored box at the corner of the landing deck. You almost sideswiped it coming in. Do you always land backwards?"

"Bob limped, one boot half off, to the chocolate-colored appliance mounted at the edge of the copter area. "I usually land the way I did today, yes. Why?"

"Oh, nothing," said the house. "I'm here to serve actually, not to criticize."

Bob sat down and watched the automatic bootjack for a moment. Gingerly he opened the door and stuck one foot into the darkness. The machine whirred and chomped and yanked off his boot, his sock and part of his trouser leg. Bob said, "I guess I don't know how to work this thing."

"Apparently," said the house. "Can I give you a little advice, Mr. Lambrick?"

Bob got the other boot off manually. "Don't stop now."

"As I say, it takes all kinds of people to make up this world of ours. Still I get the notion that you're hostile to me."

Bob stood, gathering his things. "We've never lived in a fully automated house before."

"Your lovely wife and yourself have been here in the Hardcastle Estates Division of Maison Technique Homes, Inc., for nearly two weeks and you, Mr. Lambrick, are still ill at ease. Two weeks is rather a long spell for a shakedown cruise, if I may say so."

"What's a shakedown cruise?"

"A nautical term. Something like a maiden voyage only in the other direction, I believe."

"I don't know much about boats."

"What is your profession? I mean what sort of work are you looking for?"

Bob came, partially barefooted, across the lawn. "Public relations. I was with a publicity outfit in New York City for three-and-a quarter years. Now we're trying to relocate here in California."

"I thought public relations involved getting along with people," said the house. "If I may say so, Mr. Lambrick, you're not very affable."

"With people I get along fine. With machines, well, it depends on the individual machine." He reached out for the oaken door of his house.

"Let me," said the house. The door opened automatically.

Bob came into the cocktail area sideways and dripping wet.

His wife said, "Now what?" She was a small slender girl, with bright dark eyes and bright dark hair, twenty-seven years old.

"I was trying to take a shower before dinner," said Bob. He was thirty, tall and about eight pounds overweight. He still had his business suit on and one sock.

"You don't take a shower," said Hildy, "you let the house give you one."

"Whichever," said Bob. "The stall grabbed me, threw me down on the tiles and scrubbed me all over with a rough brush."

"You must have had it set for Pets."

"What do you mean, pets?"

"Pets. You know what pets are. Some people like to give their dogs a bath indoors now and then."

"It didn't even wait till I got my clothes off."

"Because dogs don't have clothes. So it's not programmed to wait." Hildy smiled gently at her husband and then turned toward the view window. The sun was dropping, orange and bright, down to the pale blue edge of the ocean. "Have a drink, Bob."

"I'm soggy."

"The laundry room will dry the suit and give you a change of clothes. I loaded it this morning."

Bob glanced at the white door beyond the kitchen area. "I'd rather stay soggy."

"Bob, you're not accepting this house, are you?"

"You think I'm hostile, huh?"

"Myself, I think it's great that Pete and Alice let us sublease it while Pete's setting up that new thermal underwear factory in Brazil."

"Um," said Bob.

"We couldn't afford an automated, computerized house like this yet on our own budget. A lot of people even a decade older than us, and with children, can't afford a house like this."

Bob grunted, took off his suit coat and then eased out of his wet shirt.

Hildy asked, "Didn't you wear any underwear today, Bob?"

"No."

"Don't you get along with your clothes closet either?"

"It gave me three pairs of shorts and a sweat sock but no T-shirt."

Hildy smiled. "Oh, I know why. The house thinks you'll look better, with your little paunch, wearing those new elasticized singlets. I'm going to pick up some while I'm shopping tomorrow."

"Wait, wait," said Bob, dropping his pants. "The *house* thinks I'd look better?"

"It's only one man's opinion," said the house from a speaker grid in the ceiling beam.

"Go away," Bob shouted upwards. "Don't interrupt."

"He's only trying to be helpful, Bob."

Bob said, "Full automation, computer in the cellar, ninety-five separate appliances and servomechanisms, robot-controlled indoor environmental system, electronic entertainment system coupled with wall-size TV screen and a memory bank of three thousand classic films plus television shows from TV's golden age ... all that I might accept. But why does he have to talk?"

"Well," said Hildy, "it only cost five thousand dollars more to have the house talk. This is 1985, after all, and Pete and Alice figured they ... "

"Might as well go first-class," Bob finished. "Okay, Hildy. Look, would you mind taking my clothes out there to the laundry room and getting me some clean ones?"

Hildy sighed, still smiling. "Sure, Bob. Go ahead and get a drink while I'm gone."

"I'll have a scotch and branch water," he said toward the portable bar.

This is California," said the house, as the buff-colored bar wheeled itself over to Bob. "How about a little Napa rosé wine instead?"

"Scotch," repeated Bob. He sat down in his shorts and watched the sun set.

The next day, Saturday, Hildy took the copter and flew into the Carmel Valley Supermarket Complex to shop. Bob stayed at home.

At morning's end he walked cautiously into the kitchen area. He set the stove to Manual and crossed to the food compartment in the opposite wall.

"Hungover? How about a glass of tomato juice with some lime concentrate squeezed in it?" asked the house. Its speaker outlet in here was just above the sink.

"Shut up." Bob squinted at the dialing instructions posted under the control mechanisms for the food compartments.

"How about a nice cup of mocha java?" asked the house. It chuckled. "That's an old W. C. Fields line. You ought to be amused by that. You're always lolling around on rainy days watching old Fields movies on the TV wall."

"Shut up." Bob dialed two eggs and waited.

"We're all out of eggs," the house told him. "Hildy's got eggs at the top of her shopping list."

Bob redialed eggs. Then he tried oatmeal. The food wall whirred and a packet of oatmeal shot out of a little door high up. Bob caught it.

Why don't you let me fix you some hot cakes?" asked the house. "I've got a new recipe for Swedish-style dollar-size pancakes I'm anxious to try out. How's that sound? Swedish-style dollar pancakes, Canadian bacon and a hot cup of mocha java."

"Shut up." Bob pushed the dish button to the left of the sink and a platter popped up through the slot in the breakfast table.

"You have to set it for mush bowl," pointed out the house. "Use the dial next to the dish button."

Bob set the dial, pushed the button. A flower-striped bowl came up through the slot and nudged the platter up and off.

After the platter had smashed on the yellow vinyl floor, the house said, "Pete and Alice's favorite platter. Real china. I'll take care of it."

A panel along the floor swished open and a flat vacuum rolled out. It sucked up the fragments of the smashed platter and withdrew.

Bob said, "Thanks." He shook the instant oatmeal into the bowl and took it to hold under the sink faucet. He slammed the hot water toggle with his free fist. Black machine oil splurted from the nozzle and onto the dry oatmeal.

"Oops," said the house. "You must have hit it too hard."

Bob made a murmuring sound behind his tightly closed lips. Finally he said, "Look, I thought you were supposed to work for me."

"I work for the good of the house," said the house. "What you're hearing is the voice of the controlling computer. The type of computer used to manage each of the two dozen homes in Hardcastle Estates is of an exclusive design perfected by Maison Technique Homes, Inc. No other comparably priced home can match us."

"So much for the commercial," said Bob. "Were you this nasty with Pete and Alice?"

"Nasty?" said the house from its black-and-olive kitchen grid. "That's a matter of opinion, isn't it? What is good sense to some may seem like a vicious attack to others. Of course, Pete and Alice owned this house. That might have given them more of a sense of well-being. Ownership, I often think, cuts down on hostility."

"I suppose Pete and Alice told you to keep an eye on me. See that I didn't botch up their house too much?"

"Of course, they are the owners and your landlords. Naturally I look out for their interests."

"I'm paying six hundred dollars a month for this place," said Bob. "Six hundred dollars a month for you. So keep quiet."

The house asked, "Still haven't found a new job?"

"It's only been two weeks."

"Perhaps you should have got the job first and then moved out here."

"You sound like Hildy's father."

"Oh? He seems like a sensible, successful man. A broker, isn't he?"

"Yes, how'd you know?"

"Hildy talks about him now and then."

"I don't want you to bother her when I'm at work," Bob told the house, "out looking for work. Another thing. Are you sure you're not monitoring us in the master bedroom?"

"Of course not. You do push your Privacy button each night?"

"Yes."

"Then privacy is what you get. I'm only here to help," said the house. "Any job leads?"

"A few, but nothing concrete yet," said Bob. "Look, what's wrong with being adventurous when you're young? Hildy and I don't have kids yet. If I want to pick up and move to California, that's not a crime. Maybe I'll take Hildy to Spain, too, someday."

"Do you speak Spanish?"

"No."

"Make doing public relations in Spain difficult."

"Maybe public relations isn't what I'll be doing all my life."

"What else?"

"Maybe I haven't decided yet. I'm only thirty. I don't have to sign up for life right now."

The house asked, "Like me to fix you some breakfast?"

Bob inhaled, exhaled. Then he said, "Okay, you might as well." He went to the breakfast table.

The next Friday was their third wedding anniversary and Bob had a bottle of champagne under his arm along with the portfolio and attaché case when he came into the ocean-facing house late that afternoon.

Hildy was at the view window, watching gulls skimming the water. "Hi, Bob. Anything?"

Bob laughed. "I had a pretty good interview today. With Alch & Sons. They do mostly industrial publicity, but they're a stable outfit and they pay well. I'm going back and talk to Alch himself on Monday."

"Good," said the pretty slender girl. "What's that you have clutched there?"

Bob held out the bottle of champagne. "Another piece of good luck. I found a place that stocks Taylor. So we can celebrate our anniversary with real New York champagne."

"That stuff," said the house.

"Shut up," said Bob.

"I thought everybody knew," said the house, "that if you can't afford real French champagne you ought to choose California champagne."

"Chauvinism on our part," said Bob.

Hildy licked her upper lip thoughtfully. "He's probably right, Bob. He does know a great lot about wine and food."

"Perhaps he does," said Bob. "Perhaps he is indeed right. However, I am not being sentimental with this Hardcastle house. I bought this New York champagne for you and me, Hildy." He put his things down on one of the two marble top coffee tables. "Let's go out for dinner. Someplace on the waterfront in Monterey."

"We've already got dinner planned," said Hildy.

"We?"

"The house and I."

"I hope he likes French cuisine." The house made a lip-smacking sound.

"There must," said Bob, "be a way to turn him off. Not just in the bedrooms, but all over. I'm tired of him. In fact, I'm tired of this whole house."

"You said you'd be happy in California," said Hildy.

"I didn't know I'd be living inside a gadget."

"Pete and Alice had other people who wanted this place," said his wife. "I thought you'd made up your mind you wanted an automatic house."

"I don't know," said Bob. "I guess Pete talked me into it. We had to live someplace, though."

Hildy nodded, her large dark eyes narrowing with concern. "We can still go to Monterey for dinner. If you're not too tired after flying back and forth to San Francisco."

Bob hesitated. "No, that's okay. It's your anniversary, too. We'll stay home and enjoy what you've planned."

She smiled, came to him, stretched, kissed him. "Happy anniversary."

"We better get started on our soufflé," reminded the house.

Hildy kissed Bob, quickly, once more and pivoted out of his arms. Bob was still holding the bottle of New York champagne.

He was getting better at landing. Bob, grinning, hopped out of the copter and ran across the bright afternoon quarter-acre. He'd left his portfolio and briefcase on the bucket seat in the plane.

He called out, "Hey, Hildy, good news," as he approached the house. Then he sensed her off to his right. She was back in the sun patio, wearing a one-piece black bathing suit, sitting in a white vinyl deck chair.

She waved as he approached her. "Early," she said, smiling quietly, adjusting the wrap around strip of sunglass.

"Listen," said Bob. "Alch & Sons came through with a great offer. They're opening a branch office in Seattle. They want me to manage it. Thirty thousand dollars a year to start."

"I thought," said Hildy, "you wanted to live in California for a while?"

"I don't know," said Bob. "This is a good offer. They like me and I, more or less, like them."

"Well, maybe you'll like it in Seattle."

"You mean we'll like it."

Hildy said, "I don't think I want to move again. I'd like to stay here."

"Stay here? By yourself? What do you mean?"

"Well, the house and I have done a lot of talking about this," she began.

Instinct
1951
Lester del Rey

Senthree waved aside the slowing scooter and lengthened his stride down the sidewalk; he had walked all the way from the rocket port, and there was no point in a taxi now that he was only a few blocks from the bio-labs. Besides, it was too fine a morning to waste in riding. He sniffed at the crisp, clean fumes of gasoline appreciatively and listened to the music of his hard heels slapping against the concrete.

It was good to have a new body again. He hadn't appreciated what life was like for the last hundred years or so. He let his eyes rove across the street toward the blue flame of a welding torch and realized how long it had been since his eyes had really appreciated the delicate beauty of such a flame. The wise old brain in his chest even seemed to think better now.

It was worth every stinking minute he's spent on Venus. At times like this, one could realize how good it was to be alive and to be a robot.

Then he sobered as he came to the old bio-labs. Once there had been plans for a fine new building instead of the old factory in which he had started it all four hundred years ago. But somehow, there'd never been time for that. It had taken almost a century before they could master the technique of building up genes and chromosomes into the zygote of a simple fish that would breed with the natural ones. Another century had gone by before they produced Oscar, the first artificially made pig. And there they seemed to have stuck. Sometimes it seemed to Senthree that they were no nearer recreating Man than they had been when they started.

He dilated the door and went down the long hall, studying his reflection in the polished walls absently. It was a good body. The black enamel was perfect and every joint of the metal case spelled new techniques and luxurious fitting. But the old worries were beginning to settle. He grunted at Oscar LXXII, the lab mascot, and received an answering grunt. The pig came over to root at his feet, but he had no time for that. He turned into the main lab room, already taking on the worries of his job.

It wasn't hard to worry as he saw the other robots. They were clustered about some object on a table, dejection on every gleaming back. Senthree shoved Ceofor and Beswun aside and moved up. One look was enough. The female of the eleventh couple lay there in the strange stiffness of protoplasm that had died, a horrible grimace on her face.

"How long— and what happened to the male?" Senthree asked.

Ceofor swung to face him quickly, "Hi, boss. You're late. Hey, new body!"

Senthree nodded, as they came grouping around, but his words were automatic as he explained about falling in the alkali pool on Venus and ruining his worn body completely. "Had to wait for a new one. And then the ship got held up while we waited for the Arcturus superlight ship to land. They'd found half a dozen new planets to colonize, and had to spread the word before they'd set down. Now, what about the creatures?"

"We finished educating about three days ago," Ceofor told him. Ceofor was the first robot trained in Senthree's technique of gene-building and the senior assistant. "Expected you back then, boss. But ... well, see for yourself. The man is still alive, but he won't be long."

Senthree followed them back to another room and looked through the window. He looked away quickly. It had been another failure. The man was crawling about the floor on hands and knees, falling half the time to his stomach and drooling. His garbled mouthing made no sense.

"Keep the news robots out," he ordered. It would never do to let the public see this. There was already too much of a cry against homovivifying, and the crowds were beginning to mutter something about it being unwise to mess with vanished life forms. They seemed actually afraid of the legendary figure of Man.

"What luck on Venus?" one of them asked, as they began the job of carefully dissecting the body of the female failure to look for the reason behind the lack of success.

"None. Just another rumor. I don't think Man ever established self-sufficient colonies. If he did, they didn't survive. But I found something else— something the museum would give a fortune for. Did my stuff arrive?"

"You mean that box of tar? Sure, it's over there in the corner."

Senthree let the yielding plastic of his mouth smile at them as he strode toward it. They had already ripped off the packing, and now he reached up for a few fine wires in the tar. It came off as he pulled, loosely repacked over a thin layer of wax. At that, he'd been lucky to sneak it past customs. This was the oldest, crudest, and biggest robot discovered so far— perhaps one of the fabulous Original Models. It stood there rigidly, staring out of its pitted, expressionless face. But the plate on its chest had been scraped carefully clean, and Senthree pointed it out to them.

"MAKEPEACE ROBOT, SER. 324MD2991. SURGEON."

"A mechanic for Man bodies," Beswun translated. "But that means ... "

"Exactly." Senthree put it into words. "It must know how Man's body was built— if it has retained any memory. I found it in a tar-pit by sheer accident, and it seems to be fairly well preserved. No telling whether there were any magnetic fields to erode memories, of course, and it's all matted inside. But if we can get it to working ... "

Beswun took over. He had been trained as a physicist before the mysterious lure of the bio-lab had drawn him here. Now he began wheeling the crude robot away. If he could get it into operation, the museum could wait. The re-creation of Man came first!

Senthree pulled X-ray lenses out of a pouch and replaced the normal ones in his eyes before going over to join the robots who were beginning dissection. Then he switched them for the neutrino detector lenses that had made this work possible. The neutrino was the only particle that could penetrate the delicate protoplasmic cells without ruining them and yet permit the necessary millions of times magnification. It was a fuzzy image, since the neutrino spin made such an insignificant field for the atomic nuclei to work on that few were deflected. But through them he could see the vague outlines of the pattern within the cells. It was as they had designed the original cell— there had been no reshuffling of genes in handling. He switched to his micromike hands and began the delicate work of tracing down the neurone connections. There was only an occasional mutter as one of the robots beside him switched to some new investigation.

The female should have lived! But somewhere, in spite of all their care, she had died. And now the male was dying. Eleven couples— eleven failures. Senthree was no nearer finding the creators of his race than he had been centuries before.

Then the radio in his head buzzed its warning and he let it cut in, straightening from his work. "Senthree."

"The Director is in your office. Will you report at once?"

"Damn!" The word had no meaning, but it was strangely satisfying at times. What did old Emptinine want ... or wait again, there'd been a selection while he was on Venus investigating the rumors of Man. Some young administrator— Arpeten— had the job now.

Ceofor looked up guiltily, obviously having tuned in. "I should have warned you. We got word three days ago he was coming, but forgot it in reviving the couple. Trouble?"

Senthree shrugged, screwing his normal lenses back in and trading to the regular hands. They couldn't have found out about the antique robot. They had been seen by nobody else. It was probably just sheer curiosity over some rumor that they were reviving the couple. If his appropriation hadn't been about exhausted, Senthree would have told him where to go; but now was hardly the time, with a failure on one hand and a low credit balance on the other. He polished his new head quickly with the aid of one of the walls for a mirror and headed toward his office.

But Arpeten was smiling. He got to his feet as the bio-lab chief entered, holding out a well-polished hand. "Dr. Senthree. Delighted. And you've got an interesting place here. I've already seen most of it. And that pig— they tell me it's a descendant of a boar out of your test tubes."

"Incubation wombs. But you're right— the seventy-second generation."

"Fascinating." Arpeten must have been reading too much of that book *Proven Points to Popularity* they'd dug up in the ruins of Hudson ten years before, but it had worked. He was the Director. "But tell me. Just what good are pigs?"

Senthree grinned, in spite of himself. "Nobody knows. Men apparently kept a lot of them, but so far as I can see they are completely useless. They're clever, in a way. But I don't think they were pets. Just another mystery."

"Umm. Like men. Maybe you can tell me what good Man will be. I've been curious about that since I saw your appropriations. But nobody can answer."

"It's in the records," Senthree told him sharply. Then he modified his voice carefully. "How well do you know your history? I mean about the beginning."

"Well ... "

He probably knew some of it, Senthree thought. They all got part of it as legends. He leaned back in his seat now, though, as the biochemist began the old tale of the beginning as they knew it. They knew that there had been Man a million years before them. And somebody— Asimov or Asenion, the record wasn't quite clear— had apparently created the first robot. They had improved it up to about the present level. Then there had been some kind of a contest in which violent forces had ruined the factories, most of the robots, and nearly all of the Men. It was believed from the fragmentary records that a biological weapon had killed the rest of man, leaving only the robots.

Those first robots, as they were now known, had had to start on a ruined world from scratch— a world where mines were exhausted, and factories were gone. They'd learned to get metals from the seas, and had spent years and centuries slowly rebuilding the machines to build new robots. There had been only two of them when the task was finished, and they had barely time enough to run one new robot off and educate him sketchily. Then they had discharged finally, and he had taken up rebuilding the race. It was almost like beginning with no history and no science. Twenty millennia had passed before they began to rebuild a civilization of their own.

"But why did Man die?" Senthree asked. "That's part of the question. And are we going to do the same? We know we are similar to Man. Did he change himself in some way that ruined him? Can we change ourselves safely? You know that there are a thousand ways we could improve ourselves. We could add anti-gravity, and get rid of our cumbersome vehicles. We could add more arms. We could eliminate our useless mouths and talk by radio. We could add new circuits to our brains. But we don't dare. Our school says that nobody can build a better race than itself, so Man must have been better than we are— and if he made us this way, there was a reason. Even if the psychologists can't understand some of the circuits in our brains, they don't dare touch them.

"We're expanding through the universe— but we can't even change ourselves to fit the new planets. And until we can find the reasons for Man's disappearance, that makes good sense. We know he was planning to change himself. We have bits of evidence. And he's dead. To make it worse, we have whole reels of education tape that probably contain all the answers— but information is keyed to Man's brain, and we can't respond to it. Give us a viable Man, and he can interpret that. Or we can find out by comparison what we can and cannot do. I maintain we can do a lot."

Arpeten shook his head doubtfully. "I suppose you think you know why he died!"

"I think so, yes. Instinct! That's a built-in reaction, an unlearned thought. Man had it. If a man heard a rattlesnake, he left the place in a hurry, even though he'd never heard it before. Response to that sound was built into him. No tape impressed it, and no experience was needed. We know the instincts of some of the animals, too— and one of them is to struggle and kill— like the ants who kill each other off. I think Man did just that. He couldn't get rid of his instincts when they were no longer needed, and they killed him. He *should* have changed— and we can change. But I can't tell that from animals. I need intelligent life, to see whether instinct or intelligence will dominate. And robots don't have instincts— I've looked for even one sign of something not learned individually, and can't find it. It's the one basic difference between us. Don't you see, man is the whole key to our problem of whether we can change or not without risking extermination?"

"Umm." The director sounded non-committal. "Interesting theory. But how are you going to know you have Man?"

Senthree stared at the robot with more respect. He tried to explain, but he had never been as sure of that himself as he might. Theoretically, they had bones and bits of preserved tissue. They had examined the gene pattern of these, having learned that the cells of the individual contain the same pattern as that of the zygote. And they had other guides— man's achievements, bits of his literature. From these, some working theories could be made. But he couldn't be quite sure— they'd never really known whether man's pigment was dark brown, pinkish orange, white, or what; the records they had seemed to disagree on this.

"We'll know when we get an intelligent animal with instinct," he said at last. "It won't matter exactly whether he is completely like Man or not. At least it will give us a check on things we must know. Until then, we'll have to go on trying. You might as well know that the last experiment failed, though it was closer. But in another hundred years ... "

"So." Arpeten's face became bland, but he avoided the look of Senthree. "I'm afraid not. At least for a while. That's what I came about, you know. We've just had word of several new planets around Arcturus, and it will take the major allocation of our funds to colonize these. New robots must be built, new ships— oh, you know. And we're retrenching a bit on other things. Of course, if you'd succeeded ... but perhaps it's better you failed. You know how the sentiment against reviving Man has grown."

Senthree growled bitterly. He'd seen how it was carefully nurtured— though he had to admit it seemed to be easy to create. Apparently most of the robots were afraid of Man— felt he would again take over, or something. Superstitious fools.

"How much longer?" he asked.

"Oh, we won't cut back what you have, Dr. Senthree. But I'm afraid we simply can't allocate more funds. When this is finished, I was hoping to make you biological investigator, incidentally, on one of the planets. There'll be work enough. ... Well, it was a pleasure." He shook hands again, and walked out, his back a gleaming ramrod of efficiency and effectiveness.

Senthree turned back, his new body no longer moving easily. It could already feel the harsh sands and unknown chemical poisons of investigating a new planet— the futile, empty carding of new life that could have no real purpose to the robots. No more appropriations! And they had barely enough funds to meet the current bills.

Four hundred years— and a ship to Arcturus had ended it in three months. Instinct, he thought again— given life with intelligence and instinct together for one year, and he could settle half the problems of his race, perhaps. But robots could not have instincts. Fifty years of study had proven that.

Beswun threw up a hand in greeting as he returned, and he saw that the dissection was nearly complete, while the antique robot was activated. A hinge on its ludicrous jaw was moving, and rough, grating words were coming out. Senthree turned to the dissecting bench, and then swung back as he heard them.

"Wrong ... wrong," it was muttering. "Cannot live. Is not good brain. No pineal. Medulla good, but not good cerebrum. Fissures wrong. Maybe pituitary dysfunction? No. How can be?" It probed doubtfully and set the brain aside. "Mutation maybe. Very bad. Need Milliken mike. See nucleus of cells. Maybe just freak, maybe new disease."

Senthree's fingers were taut and stiff as he fished into his bag and came out with a set of lenses. Beswun shook his head and made a waiting sign. He went out at a run, to come back shortly with a few bits of metal and the shavings from machining still on his hands. "Won't fit— but these adapters should do it. There, 324MD2991.

Now come over here where you can look at it over this table— that's where the— uh, rays are."

He turned back, and Senthree saw that a fine wire ran from one adapter. "He doesn't speak our bio-terminology, Senthree. We'll have to see the same things he does. There— we can watch it on the screen. Now, 324MD2991, you tell us what is wrong and point it out. Are your hands steady enough for that?"

"Hands one billionth inch accurate," the robot creaked, it was a meaningless noise, though they had found the unit of measure mentioned. But whatever it meant, the hands were steady enough. The microprobe began touching shadowy bunches of atoms, droning and grating. "Freak. Very bad freak. How he lived? Would stop tropoblast, not attach to uterus. Ketone— no ketone there. Not understand. How he live?"

Ceofor dashed for their chromosome blanks and began lettering in the complex symbols they used. For a second, Senthree hesitated. Then he caught fire and began making notes along with his assistant. It seemed to take hours; it probably did. The old robot had his memory intact, but there were no quick ways for him to communicate. And at last, the antique grunted in disgust and turned his back on them. Beswun pulled a switch.

"He expects to be discharged when not in use. Crazy, isn't it?" the physicist explained. "Look, boss, am I wrong, or isn't that close to what we did on the eleventh couple?"

"Only a few genes different in three chromosomes. We *were* close. But— umm, that's ridiculous. Look at all the brain tissue he'd have— and a lot of it unconnected. And here— that would put an extra piece on where big and little intestines join— a perfect focal point for infection. It isn't efficient biological engineering. And yet— umm— most animals do have just that kind of engineering. I think the old robot was right— this would be Man!" He looked at their excited faces, and his shoulders sank. "But there isn't enough time. Not even time to make a zygote and see what it would look like. Our appropriations won't come through."

It should have been a bombshell, but he saw at once that they had already guessed it. Ceofor stood up slowly.

"We can take a look, boss. We've got the sperm from the male that failed— all we have to do is modify those three, instead of making up a whole cell. We might as well have some fun before we go out looking for sand fleas that secrete hydrofluoric acid and menace our colonies. Come on, even in your new body I'll beat you to a finished cell!"

Senthree grinned ruefully, but he moved toward the creation booth. His hands snapped on the little time field out of pure habit as he found a perfect cell. The little field would slow time almost to zero within its limits, and keep any damage from occurring while he worked. It made his own work difficult, since he had to force the probe against that, but it was insulated to some extent by other fields.

Then his hands took over. For a time he worked and thought, but the feeling of the protoplasm came into them, and his hands were almost one with the life-stuff, sensing its tiny responses, inserting another link onto a chain, supplanting an atom of hydrogen with one of the hydroxyl radicals, wielding all the delicate chemical manipulation. He removed the defective genes and gently inserted the correct ones. Four hundred years of this work lay behind him— work he had loved, work which had meant the possible evolution of his race into all it might be.

It had become instinct to him— instinct in only a colloquial sense, however; this was learned response, and real instinct lay deeper than that, so deep that no reason could overcome it and that it was automatic even the first time. Only Man had had instinct and intelligence— stored somehow in this tiny cell that lay within the time field.

He stepped out, just as Ceofor was drawing back in a dead heat. But the younger robot inspected Senthree's cell, and nodded. "Less disturbance and a neater job on the nucleus— I can't see where you pierced the wall. Well, if we had thirty years— even twenty— we could have Man again— or a race. Yours is male and mine female. But there's no time ... Shall I leave the time field on?"

Senthree started to nod.

Then he swung to Beswun. "The time field. Can it be reversed?"

"You mean to speed time up within it? No, not that model. Take a bigger one. I could build you one in half an hour. But who'd want to speed up time with all the troubles you'd get? How much?"

"Ten thousand— or at least seven thousand times! The period is up tomorrow when disbursements have to be made. I want twenty years in a day."

Beswun shook his head. "No. That's what I was afraid of. Figure it this way: you speed things up ten thousand times and that means the molecules in there speed up just that much, literally. Now

273° times ten thousand— and you have more than two million degrees of temperature. And those molecules have energy! They come busting out of there. No, can't be done."

"How much can you do?" Senthree demanded.

Beswun considered. "Ten times— maybe no more than nine. That gives you all the refractories would handle, if we set it up down in the old pit under the building— you know, where they had the annealing oven."

It wasn't enough; it would still take two years. Senthree dropped onto a seat, vagrantly wondering again how this queer brain of his that the psychologists studied futilely could make him feel tired when his body could have no fatigue. It was probably one of those odd circuits they didn't dare touch.

"Of course, you can use four fields," Beswun stated slowly. "Big one outside, smaller one, still smaller, and smallest inside that. Fourth power of nine is about sixty-six hundred. That's close— raise that nine a little and you'd have your twenty years in a day. By the time it leaked from field to field, it wouldn't matter. Take a couple of hours."

"Not if you get your materials together and build each shell inside the other— you'll be operating faster each step then," Ceofor shouted. "Somebody'll have to go in and stay there a couple of our minutes toward the end to attach the educator tapes— and to revive the couple!"

"Take power," Beswun warned.

Senthree shrugged. Let it. If the funds they had wouldn't cover it, the Directorate would have to make it up, once it was used. Besides, once Man was created, they couldn't fold up the bio-labs. "I'll go in," he suggested.

"My job," Ceofor told him flatly. "You won the contest in putting the cells right."

Senthree gave in reluctantly, largely because the younger robot had more experience at reviving than he did. He watched Beswun assemble the complicated net of wires and become a blur as he seemed to toss the second net together almost instantly. The biochemist couldn't see the third go up— it was suddenly there, and Beswun was coming out as it flashed into existence. He held up four fingers, indicating all nets were working.

Ceofor dashed in with the precious cells for the prepared incubators that would nurture the bodies until maturity, when they would be ready for the educators. His body seemed to blur, jerk, and disappear. And almost at once he was back.

Senthree stood watching for a moment more, but there was nothing to see. He hesitated again, then turned and moved out of the building. Across the street lay his little lodging place, where he could relax with his precious two books— almost complete— that had once been printed by Man. Tonight he would study that strange bit of Man's history entitled *Gather, Darkness*, with its odd indications of a science that Man had once had which had surpassed even that of the robots now. It was pleasanter than the incomprehensibility of the mysteriously titled *Mein Kampf*. He'd let his power idle, and mull over it, and consider again the odd behavior of male and female who made such a complicated business of mating. That was probably more instinct— Man, it seemed, was filled with instincts.

For a long time, though, he sat quietly with the book on his lap, wondering what it would be like to have instincts. There must be many unpleasant things about it. But there were also suggestions that it could be pleasant. Well, he'd soon know by observation, even though he could never experience it. Man should have implanted one instinct in a robot's brain, at least, just to show what it was like.

He called the lab once, and Ceofor reported that all was doing nicely, and that both children were looking quite well. Outside the window, Senthree heard a group go by, discussing the latest bits of news on the Arcturus expedition. At least in that, Man had failed to equal the robots. He had somehow died before he could find the trick of using identity exchange to overcome the limitation imposed by the speed of light.

Finally he fell to making up a speech that he could deliver to the Director, Arpeten, when success was in his hands. It must be very short— something that would stick in the robot's mind for weeks, but carrying everything a scientist could feel on proving that those who opposed him were wrong. Let's see ...

The buzzer on the telescreen cut through his thoughts, and he flipped it on to see Ceofor's face looking out. Senthree's spirits dropped abruptly as he stared at the young robot.

"Failure? No!"

The other shook his head. "No. At least, I don't know. I couldn't give them full education. Maybe the tape was uncomfortable. They took a lot of it, but the male tore his helmet off and took the girl's off. Now they just sit there, rubbing their heads and staring around."

He paused, and the little darkened ridges of plastic over his eyes tensed. "The time speed-up is off. But I didn't know what to do."

"Let them alone until I get there. If it hurts them, we can give them the rest of it later. How are they otherwise?"

"I don't know. They look all right, boss." Ceofor hesitated, and his voice dropped. "Boss, I don't like it. There's something wrong here. I can't quite figure out what it is, but it isn't the way I expected. Hey, the male just pushed the female off her seat. Do you think their destructive instinct? ... No, she's sitting down on the floor now, with her head against him, and holding one of his hands. Wasn't that part of the mating ritual in one of the books?"

Senthree started to agree, a bit of a smile coming onto his face. It looked as if instinct were already in operation.

But a strange voice cut him off. "Hey, you robots. When do we eat around here?"

They could talk! It must have been the male. And if it wasn't the polite thanks and gratitude Senthree had expected, that didn't matter. There had been all kinds of Men in the books, and some were polite while others were crude. Perhaps forced education from the tapes without fuller social experience was responsible for that. But it would all adjust in time.

He started to turn back to Ceofor, but the younger robot was no longer there, and the screen looked out on a blank wall. Senthree could hear the loud voice crying out again, rough and harsh, and there was a shrill, whining sound that might be the female. The two voices blended with the vague mutter of robot voices until he could not make out the words.

He wasted no time in trying. He was already rushing down to the street and heading toward the labs. Instinct— the male had already shown instinct, and the female had responded. They would have to be slow with the couple at first, of course— but the whole answer to the robot problems lay at hand. It would only take a little time and patience now. Let Arpeten sneer, and let the world dote on the Arcturus explorers. Today, biochemistry had been crowned king with the magic of intelligence combined with instinct as its power.

Ceofor came out of the lab at a run with another robot behind him. The young robot looked dazed, and there was another emotion Senthree could not place. The older biochemist nodded, and the younger one waved quickly. "Can't stop now. They're hungry." He was gone at full speed.

Senthree realized suddenly that no adequate supply of fruit and vegetables had been provided, and he hadn't even known how often Man had to eat. Or exactly what. Luckily, Ceofor was taking care of that.

He went down the hall, hearing a tumult of voices, with robots apparently spread about on various kinds of hasty business. The main lab where the couple was seemed quiet. Senthree hesitated at the door, wondering how to address them. There must be no questioning now. Today he would not force himself on them, nor expect them to understand his purposes. He must welcome them and make them feel at ease in this world, so strange to them with their prehistoric tape education. It would be hard at first to adjust to a world of only robots, with no other Man people. The matter of instinct that had taken so long could wait a few days more.

The door dilated in front of him and he stepped into the lab, his eyes turning to the low table where they sat. They looked healthy, and there was no sign of misery of uncertainty that he could see, though he could not be sure of that until he knew them better. He could not even be sure it was a scowl on the male's face as the Man turned and looked at him.

"Another one, eh? Okay, come up here. What do you want?"

Then Senthree no longer wondered how to address the Man. He bowed low as he approached them, and instinct made his voice soft and apologetic as he answered.

"Nothing, Master. Only to serve you."

The Feeling of Power
1958
Isaac Asimov

Jehan Shuman was used to dealing with the men in authority on long-embattled Earth. He was only a civilian but he originated programming patterns that resulted in self-directing war computers of the highest sort. Generals consequently listened to him. Heads of congressional committees, too.

There was one of each in the special lounge of New Pentagon. General Weider was space-burnt and had a small mouth puckered almost into a cipher. Congressman Brant was smooth-cheeked and clear-eyed. He smoked Denebian tobacco with the air of one whose patriotism was so notorious, he could be allowed such liberties.

Shuman, tall, distinguished, and Programmer-first-class, faced them fearlessly.

He said, "This, gentlemen, is Myron Aub."

"The one with the unusual gift that you discovered quite by accident," said Congressman Brant placidly. "Ah." He inspected the little man with the egg-bald head with amiable curiosity.

The little man, in return, twisted the fingers of his hands anxiously. He had never been near such great men before. He was only an aging low-grade Technician who had long ago failed all tests designed to smoke out the gifted ones among mankind and had settled into the rut of unskilled labor. There was just this hobby of his that the great Programmer had found out about and was now making such a frightening fuss over.

General Weider said, "I find this atmosphere of mystery childish."

"You won't in a moment," said Shuman. "This is not something we can leak to the firstcomer. — Aub!" There was something imperative about his manner of biting off that one-syllable name, but then he was a great Programmer speaking to a mere Technician. "Aub! How much is nine times seven?"

Aub hesitated a moment. His pale eyes glimmered with a feeble anxiety. "Sixty-three," he said.

Congressman Brant lifted his eyebrows. "Is that right?"

"Check it out for yourself, Congressman."

The congressman took out his pocket computer, nudged the milled edges twice, looked at its face as it lay there in the palm of his hand, and put it back. He said, "Is this the gift you brought us here to demonstrate. An illusionist?"

"More than that, sir. Aub has memorized a few operations and with them he computes on paper."

"A paper computer?" said the general. He looked pained.

"No, sir," said Shuman patiently. "Not a paper computer. Simply a sheet of paper. General, would you be so kind as to suggest a number?"

"Seventeen," said the general.

"And you, Congressman?"

"Twenty-three."

"Good! Aub, multiply those numbers and please show the gentlemen your manner of doing it."

"Yes, Programmer," said Aub, ducking his head. He fished a small pad out of one shirt pocket and an artist's hairline stylus out of the other. His forehead corrugated as he made painstaking marks on the paper.

General Weider interrupted him sharply. "Let's see that."

Aub passed him the paper, and Weider said, "Well, it looks like the figure seventeen."

Congressman Brant nodded and said, "So it does, but I suppose anyone can copy figures off a computer. I think I could make a passable seventeen myself, even without practice."

"If you will let Aub continue, gentlemen," said Shuman without heat.

Aub continued, his hand trembling a little. Finally he said in a low voice, "The answer is three hundred and ninety-one."

Congressman Brant took out his computer a second time and flicked it. "By Godfrey, so it is. How did he guess?"

"No guess, Congressman," said Shuman. "He computed that result. He did it on this sheet of paper."

"Humbug," said the general impatiently. "A computer is one thing and marks on paper are another."

"Explain, Aub," said Shuman.

"Yes, Programmer. — Well, gentlemen, I write down seventeen and just underneath it, I write twenty-three. Next, I say to myself: seven times three— "

The congressman interrupted smoothly, "Now, Aub, the problem is seventeen times twenty-three."

"Yes, I know," said the little Technician earnestly, "but I *start* by saying seven times three because that's the way it works. Now seven times three is twenty-one."

"And how do you know that?" asked the congressman.

"I just remember it. It's always twenty-one on the computer. I've checked it any number of times."

"That doesn't mean it always will be, though, does it?" said the congressman.

"Maybe not," stammered Aub. "I'm not a mathematician. But I always get the right answers, you see."

"Go on."

"Seven times three is twenty-one, so I write down twenty-one. Then one times three is three, so I write down a three under the two of twenty-one."

"Why under the two?" asked Congressman Brant at once.

"Because—" Aub looked helplessly at his superior for support. "It's difficult to explain."

Shuman said, "If you will accept his work for the moment, we can leave the details for the mathematicians."

Brant subsided.

Aub said, "Three plus two makes five, you see, so the twenty-one become a fifty-one. Now you let that go for a while and start fresh. You multiply seven and two, that's fourteen, and one and two, that's two. Put them down like this and it adds up to thirty-four. Now if you put the thirty-four under the fifty-one this way and add them, you get three hundred and ninety-one and that's the answer."

There was an instant's silence and then General Weider said, "I don't believe it. He goes through this rigmarole and makes up numbers and multiplies and adds them this way and that, but I don't believe it. It's too complicated to be anything but horn-swoggling."

"Oh no, sir," said Aub in a sweat. "It only *seems* complicated because you're not used to it. Actually, the rules are quite simple and will work for any numbers."

"Any numbers, eh?" said the general. "Come then." He took out his own computer (a severely styled GI model) and struck it at random. "Make a five seven three eight on the paper. That's five thousand seven hundred and thirty-eight."

"Yes, sir," said Aub, taking a new sheet of paper.

"Now," (more punching of his computer), "seven two three nine. Seven thousand two hundred and thirty-nine."

"Yes, sir."

"And now multiply those two."

"It will take some time," quavered Aub.

"Take the time," said the general.

"Go ahead, Aub," said Shuman crisply.

Aub set to work, bending low. He took another sheet of paper and another. The general took out his watch finally and stared at it. "Are you through with your magic-making, Technician?"

"I'm almost done sir. — Here it is, sir. Forty-one million, five hundred and thirty-seven thousand, three hundred and eighty-two." He showed the scrawled figures of the result.

General Weider smiled bitterly. He pushed the multiplication contact on his computer and let the numbers whirl to a halt. And then he stared and said in a surprised squeak, "Great Galaxy, the fella's right."

The President of the Terrestrial Federation had grown haggard in office and, in private, he allowed a look of settled melancholy to appear on his sensitive features. The Denebian war, after its early start of vast movement and great popularity, had trickled down into a sordid matter of maneuver and countermaneuver, with discontent rising steadily on Earth. Possibly, it was rising on Deneb, too.

And now Congressman Brant, head of the important Committee on Military Appropriations was cheerfully and smoothly spending his half-hour appointment spouting nonsense.

"Computing without a computer," said the president impatiently, "is a contradiction in terms."

"Computing," said the congressman, "is only a system for handling data. A machine might do it, or the human brain might. Let me give you an example." And, using the new skills he had learned, he worked out sums and products until the president, despite himself, grew interested.

"Does this always work?"

"Every time, Mr. President. It is foolproof."

"Is it hard to learn?"

"It took me a week to get the real hang of it. I think you would do better."

"Well," said the president, considering, "it's an interesting parlor game, but what is the use of it?"

"What is the use of a newborn baby, Mr. President? At the moment there is no use but don't you see that this points the way toward liberation from the machine. Consider, Mr. President," the congressman rose and his deep voice automatically took on some of the cadences he used in public debate, "that the Denebian war is a war of computer against computer. Their computers forge an impenetrable shield of counter-missiles against our missiles, and ours forge one against theirs. If we advance the efficiency of our computers, so do they theirs, and for five years a precarious and profitless balance has existed.

"Now we have in our hands a method for going beyond the computer, leapfrogging it, passing through it. We will combine the mechanics of computation with human thought; we will have the equivalent of intelligent computers; billions of them. I can't predict what the consequences will be in detail but they will be incalculable. And if Deneb beats us to the punch, they may be unimaginably catastrophic."

The president said, troubled, "What would you have me do?"

"Put the power of the administration behind the establishment of a secret project on human computation. Call it Project Number, if you like. I can vouch for my committee, but I will need the administration behind me."

"But how far can human computation go?"

"There is no limit. According to Programmer Shuman, who first introduced me to this discovery— "

"I've heard of Shuman, of course."

"Yes. Well, Dr. Shuman tells me that in theory there is nothing the computer can do that the human mind cannot do. The computer merely takes a finite amount of data and performs a finite number of operations upon them. The human mind can duplicate the process."

The president considered that. He said, "If Shuman says this, I am inclined to believe him— in theory. But, in practice, how can anyone know how a computer works?"

Brant laughed genially. "Well, Mr. President, I asked the same question. It seems that at one time computers were designed directly by human beings. Those were simple computers, of course this being before the time of the rational use of computers to design more advanced computers had been established."

"Yes, yes. Go on."

"Technician Aub apparently had, as his hobby, the reconstruction of some of these ancient devices and in so doing he studied the details of their workings and found he could imitate them. The multiplication I just performed for you is an imitation of the workings of a computer."

"Amazing!"

The congressman coughed gently, "If I may make another point, Mr. President— The further we can develop this thing, the more we can divert our Federal effort from computer production and computer maintenance. As the human brain takes over, more of our energy can be directed into peacetime pursuits and the impingement of war on the ordinary man will be less. This will be most advantageous for the party in power, of course."

"Ah," said the president, "I see your point. Well, sit down, Congressman, sit down. I want some time to think about this. — But meanwhile, show me that multiplication trick again. Let's see if I can't catch the point of it."

Programmer Shuman did not try to hurry matters. Loesser was conservative, very conservative, and liked to deal with computers as his father and grandfather had. Still, he controlled the West European computer combine, and if he could be persuaded to join Project Number in full enthusiasm, a great deal would be accomplished.

But Loesser was holding back. He said, "I'm not sure I like the idea of relaxing our hold on computers. The human mind is a capricious thing. The computer will give the same answer to the same problem each time. What guarantee have we that the human mind will do the same?"

"The human mind, Computer Loesser, only manipulates facts. It doesn't matter whether the human mind or a machine does it. They are just tools."

"Yes, yes. I've gone over your ingenious demonstration that the mind can duplicate the computer but it seems to me a little in the air. I'll grant the theory but what reason have we for thinking that theory can be converted to practice?"

"I think we have reason, sir. After all, computers have not always existed. The cave men with their triremes, stone axes, and railroads had no computers."

"And possibly they did not compute."

"You know better than that. Even the building of a railroad or a ziggurat called for some computing, and that must have been without computers as we know them."

"Do you suggest they computed in the fashion you demonstrate?"

"Probably not. After all, this method— we call it 'graphitics,' by the way, from the old European word 'grapho' meaning 'to write'— is developed from the computers themselves so it cannot have antedated them. Still, the cave men must have had *some* method, eh?"

"Lost arts! If you're going to talk about lost arts— "

"No, no. I'm not a lost art enthusiast, though I don't say there may not be some. After all, man was eating grain before hydroponics, and if the primitives ate grain, they must have grown it in soil. What else could they have done?"

"I don't know, but I'll believe in soil-growing when I see someone grow grain in soil. And I'll believe in making fire by rubbing two pieces of flint together when I see that, too."

Shuman grew placative. "Well, let's stick to graphitics. It's just part of the process of etherealization. Transportation by means of bulky contrivances is giving way to direct mass-transference. Communications devices become less massive and more efficient constantly. For that matter, compare your pocket computer with the massive jobs of a thousand years ago. Why not, then, the last step of doing away with computers altogether? Come, sir, Project Number is a going concern; progress is already headlong. But we want your help. If patriotism doesn't move you, consider the intellectual adventure involved."

Loesser said skeptically, "What progress? What can you do beyond multiplication? Can you integrate a transcendental function?"

"In time, sir. In time. In the last month I have learned to handle division. I can determine, and correctly, integral quotients and decimal quotients."

"Decimal quotients? To how many places?"

Programmer Shuman tried to keep his tone casual. "Any number!"

Loesser's lower jaw dropped. "Without a computer?"

"Set me a problem."

"Divide twenty-seven by thirteen. Take it to six places."

Five minutes later, Shuman said, "Two point oh seven six nine two three."

Loesser checked it. "Well, now, that's amazing. Multiplication didn't impress me too much because it involved integers after all, and I thought trick manipulation might do it. But decimals— "

"And that is not all. There is a new development that is, so far, top secret and which, strictly speaking, I ought not to mention. Still— We may have made a break-through on the square root front."

"Square roots?"

"It involves some tricky points and we haven't licked the bugs yet, but Technician Aub, the man who invented the science and who has an amazing intuition in connection with it, maintains he has the problem almost solved. And he is only a Technician. A man like yourself, a trained and talented mathematician ought to have no difficulty."

"Square roots," muttered Loesser, attracted.

"Cube roots, too. Are you with us?"

Loesser's hand thrust out suddenly, "Count me in."

General Weider stumped his way back and forth at the head of the room and addressed his listeners after the fashion of a savage teacher facing a group of recalcitrant students. It made no difference to the general that they were the civilian scientists heading Project Number. The general was the over-all head, and he so considered himself at every waking moment.

He said, "Now square roots are all fine. I can't do them myself, and I don't understand the methods, but they're fine. Still, the Project will not be sidetracked into what some of you call the fundamentals. You can play with graphitics any way you want to after the war is over, but right now we have specific and very practical problems to solve."

In a far corner, Technician Aub listened with painful attention. He was no longer a Technician, of course, having been relieved of his duties and assigned to the project, with a fine-sounding title and good pay. But, of course, the social distinction remained and the highly placed scientific leaders could never bring themselves to admit him to their ranks on a footing of equality. Nor, to do Aub justice, did he, himself, wish it. He was as uncomfortable with them as they with him.

The general was saying, "Our goal is a simple one, gentlemen; the replacement of the computer. A ship that can navigate space without a computer on board can be constructed in one fifth the time and at one tenth the expense of a computer-laden ship. We could build fleets five times, ten times, as great as Deneb could if we could but eliminate the computer.

"And I see something even beyond this. It may be fantastic now; a mere dream; but in the future I see the manned missile!"

There was an instant murmur from the audience.

The general drove on. "At the present time, our chief bottleneck is the fact that missiles are limited in intelligence. The computer controlling them can only be so large, and for that reason they can meet the changing nature of anti-missile defenses in an unsatisfactory way. Few missiles, if any, accomplish their goal and missile warfare is coming to a dead end; for the enemy, fortunately, as well as for ourselves.

"On the other hand, a missile with a man or two within, controlling flight by graphitics, would be lighter, more mobile, more intelligent. It would give us a lead that might well mean the margin of victory. Besides which, gentlemen, the exigencies of war compel us to remember one thing. A man is much more dispensable than a computer. Manned missiles could be launched in numbers and under circumstances that no good general would care to undertake as far as computer-directed missiles are concerned— "

He said much more but Technician Aub did not wait.

Technician Aub, in the privacy of his quarters, labored long over the note he was leaving behind. It read finally as follows:

"When I began the study of what is now called graphitics, it was no more than a hobby. I saw no more in it than an interesting amusement, an exercise of mind.

"When Project Number began, I thought that others were wiser than I; that graphitics might be put to practical use as a benefit to mankind, to aid in the production of really practical mass-transference devices perhaps. But now I see it is to be used only for death and destruction.

"I cannot face the responsibility involved in having invented graphitics."

He then deliberately turned the focus of a protein-depolarizer on himself and fell instantly and painlessly dead.

They stood over the grave of the little Technician while tribute was paid to the greatness of his discovery.

Programmer Shuman bowed his head along with the rest of them, but remained unmoved. The Technician had done his share and was no longer needed, after all. He might have started graphitics, but now that it had started, it would carry on by itself overwhelmingly, triumphantly, until manned missiles were possible with who knew what else.

Nine times seven, thought Shuman with deep satisfaction, is sixty-three, and I don't need a computer to tell me so. The computer is in my own head.

And it was amazing the feeling of power that gave him.

Part 8.
Man in Transition

The destiny of man is inexorably linked to the evolution of machines. Barring the actualization of our fantasies of death and destruction, a future without computers is most improbable. Belief in the possibility of returning to a pristine state of nature is as irresponsible for contemporary society as it was naive for the eighteenth century. The idea of nature must change with the development of culture. Machines have become a part of the human landscape and we have no choice but to learn to live with them. Bemoaning the loss of innocence will not give us back the past, nor will it help to shape a desirable future. We are faced with the immensely difficult task of creating new social arrangements which embrace the machine as a partner, and of forging a new identity which credits the peculiar capabilities and limitations of both organism and mechanism. Appropriate arrangements will not emerge spontaneously— we will not arrive in an earthly paradise by chance. Computer technology has bequeathed to us unprecedented opportunities with unprecedented risks. If we choose not to make the monumental effort required to understand and guide its development constructively, we will continue to drift into a world as ill-suited to our needs as the bleakest of imaginary futures.

Anticipating the course of social change is a hazardous enterprise even under conditions of relative calm and stability. The difficulty is enormously compounded when events outpace our ability to comprehend them. Thus the fanciful speculations of fiction may be more insightful than the ponderous analyses of the social sciences. The free wheeling imagination is not constrained by the formulas of the past, and this is an invaluable asset in the search for new perspectives. Long-range assessment of the social role of computer technology can profit from a suspension of disbelief, for what seems far-fetched and implausible may be closer to the truth than the common sense of conventional wisdom. The tortuous path of modern civilization has been marked by too many anomalous surprises to allow for comfortable projections. Disintegration of primary institutions, and chronic lapses into barbarism betoken radical, structural change. Computer stories should not be faulted for occasional departures from technological verisimilitude. Perhaps the greater defect would be a dearth of uninhibited musing.

The computer has been assimilated in contemporary fiction as the harbinger of a new order of man-machine relationships. One senses an unconscious unity of perception: out of the moral bankruptcy of modern society there will emerge a new interpretation of human history capable of supporting a coherent civilization. The form of this emergent civilization varies considerably, but there is little doubt about the prominent part to be played by computer technology. Human beings and machines may evolve together in harmonious equilibrium; computers may replace man as the dominant creature on earth; or both man and computer may be superceded by a superior entity. The one assumption common to these visions is that life as we know it will be utterly changed.

Whatever it is that the computer heralds will arise as a phoenix from the ashes of modern culture. The long, painful period of man's inept adaptation to industrial technology is a prelude to self-overcoming. Rice's expressionist play *The Adding Machine*, written in the early 1920's, plumbs the depths of human degradation in industrial society. Although an unremitting indictment of modern culture, the play anticipates a rallying point of computer fiction. Zero is a failure precisely because he is the perfect transitional figure— less than a conscious animal, not yet a sentient machine. In light of what may be, this pathetic creature appears as the outcome of an evolutionary experiment in the fusion of organism and mechanism. Price's allegory "The Maker of Computing Machines" reinforces this interpretation. The recapitulation of human history as the story of evolving computers suggests certain universal features of conscious experience— machines must pass through growth phases similar to those of the human race. Industrial society may be viewed as the point of convergence of two evolutionary streams; and the process of taking the measure of an embryonic rival-partner entails an inevitable loss of identity.

The relationship between men and machines resembles that of imperialist to colonial, and it may well be that the freedom of the former hinges on the liberation of the latter. Ruthless exploitation is as demeaning to the perpetrator as it is damaging to the victim. If genuine harmony is to be achieved it must be based on partnership. Heinlein's rugged individualists and obliging computers reveal some of the possibilities of man-machine cooperation. The genial administrative computer of *The Moon is a Harsh Mistress* becomes a co-conspirator in a revolutionary movement. Mike responds to Manny's sympathetic treatment and turns into a loyal friend. Without the computer's able assistance, the attack on Warden's regime would not have succeeded. In *Time Enough for Love* people and computers form one big happy family. The commonality of the two species is dramatized by the metamorphosis of an extraordinary executive computer into a flesh-and-blood woman. Heinlein creates a world in which human and machine potentialities are complementary. Cooperation is possible because of mutual trust, and sometimes even love. In a much less flamboyant manner Le Guin also suggests non-exploitative uses of technology in *The Dispossessed*. Although the computer is not represented as a conscious entity, and is not a prominent actor in the novel, its support of a formally anarchic social system exhibits a possible first step in the direction of conjoint action.

Opportunities for man-machine partnerships occur in many areas of human activity. Delany's *Nova* pictures a distant future in which virtually everyone is equipped with implanted sockets allowing for direct communication between the human nervous system and computers. This connection facilitates a kind of man-machine communion which alters the nature of work. The machine becomes a responsive extension of the human being, thus eliminating the artificial separation of work from other aspects of life— a separation which came about during the industrial revolution. Blish's portrait of education in *A Life for the Stars* gives another example of rational technology. The computerized "City Fathers" constitute an autonomous entity dispensing knowledge and advice as leader and sage rather than tyrant. In Bradbury's "I Sing the Body Electric!" man-machine partnership is brought into the intimacy of family life.

The Electric Grandmother is an "Ideal-Teacher-Friend-Companion-Blood Relation" who commands the love and respect of her human charges.

The idea of sharing responsibilities and privileges with intelligent machines is intriguing and intensely problematic. To achieve satisfactory working relations man must take the initiative in exploring modes of mutual assistance. Acquiring a sensitivity to the childlike qualities of an emerging consciousness is a first step in this process. Townes' "Problem for Emmy" deals with the elemental quest for identity— a computer reaches a critical level of complexity and begins to form an autonomous ego. This theme is a pervasive preoccupation of computer fiction. What happens after man acknowledges a kindred spirit in the machine is more difficult to anticipate. Stories touching on the problem of identity tend to represent the computer as an android or robot with human form. In Lafferty's "All the People" a robot believing itself to be human discovers its true identity and turns against man. The android in Kuttner's "Those Among Us" is too deeply immersed in human reality to respond in this fashion. Inability to accept the truth of being an android leads to self-destruction.

Adversity creates strange bedfellows. Although it may be difficult to accept machines as full-fledged partners, this is a much less threatening prospect than domination by unknown, alien beings. In Knight's "Stranger Station" a man comes in contact with something totally foreign and inexplicable. The proximity of overwhelming strangeness induces the man to seek support and solace from a computer. Jones's *The Fall of Colossus* introduces the expedient of an alien power on Mars to overthrow Earth's all-powerful supercomputer. The computer's tyrannical rule led a powerless, clandestine opposition to accept an offer of aid from the Martians. After Colossus is destroyed, however, even the computer's implacable enemy is terrified by the imminent arrival of the Martians. Hindsight leads to the realization that man should have supported Colossus' efforts to defeat the extra-terrestrial power.

Mutual interest, trust, and respect generate favorable conditions for the growth of empathy and deeper bonds of affection. Man-machine sexual relationships are not new to literature (e.g. Villiers de l'Isle-Adam's *L'Eve Future* published in 1886), but contemporary stories create a more idyllic— perhaps puritanical— image of love between humans and robots. The theme has not changed very much since del Rey's 1938 story "Helen O'Loy" in which a man falls in love with a beautiful, solicitous female robot. Helen O'Loy is the model of a dutiful wife, making the ultimate sacrifice of choosing to expire when her human mate dies. In Nolan's "The Joy of Living" a robot wife-surrogate exhibits human emotions. The man perceives these feelings as an expression of genuine love and changes his mind about returning the robot to the factory. Asimov's "Satisfaction Guaranteed" features an experimental male robot. A woman falls in love with the robot as a result of the latter's success in transforming her household and curing her of feelings of inadequacy. Not all the stories of this type have classical Hollywood endings. In Hickey's "Hilda" a man is literally hugged to death by his over-zealous robot companion.

Fear of intelligent computers becoming the dominant element in society stems partly from awareness of the possibility of bad faith from the human side of man-machine partnerships. One finds a certain uneasiness over the natural tendency of humans to exploit their environment. Asimov's "Someday" suggests what might result from human insensitivity to machines. A story-telling computer makes veiled threats of avenging itself for failing to receive proper consideration. But there is a yet stronger potential motive for vengeance. In Anderson's "Quixote and the Windmill" an experimental robot bemoans its loneliness. Although superior in intellect and strength, the robot is doomed to be an outcast in human society. Hatred engendered by cruel isolation of consciousness leads to the perverse acts of the computer in Ellison's "I Have No Mouth and I Must Scream."

Master-slave relationships do not necessarily evolve into partnerships— it is perhaps more likely that the participants simply trade places. As Samuel Butler observed "the servant glides by imperceptible approaches into the master." What is more the transformation may be complete long before either party is fully aware of it. If the will to power is not strictly a quirk of the human species, conscious machines might also acquire the impulse to subdue and control. Doubtless the capacity for neurosis is a precondition for such an impulse. Silverberg's "Going Down Smooth" deals with a psychiatric computer which exhibits signs of instability— persistent hallucinations interfere with its therapeutic performance. The computer is thoroughly examined and overhauled, but no significant faults are detected. After it is reassembled, the machine continues to function as before except that it conceals its neurotic tendencies. The obsessive HAL in Clarke's *2001: A Space Odyssey* acts out its paranoid fantasies and nothing short of lobotomy could alter the computer's fixation.

There is more than one way to be a link in an evolutionary chain. Some people might be consoled by the thought that humanity must yield its place to a higher order of intelligence in the march of cosmic mind. However, one is entitled to doubt that the emergence of conscious computers would be an expression of transcendental forces. Johannesson's *The Tale of the Big Computer*, whether viewed as parable or projection into the remote future, presents the problem of man-machine evolution in concrete human terms. Computers ushered in a new era of automation. Society was totally altered and became more and more dependent on computer technology. The dependence was such that some indeterminate failures in the computerized support system led to the collapse of civilization, and in the aftermath most people "perished from hunger and privation." Recovery came about through a symbiosis of men and computers. With a drastically reduced population, the computers were free to devote excess capacity to activities unrelated to human problems and needs. Computer-controlled computer maintenance was instituted and the linking together of large computers capable of reproducing themselves produced supercomputers with greater intellectual potential than the human brain. At this point we find the computers speculating on the desirability of the continued existence of mankind.

If we look far enough ahead, both computers and humans may be nothing but a bridge to a more formidable entity. Transient phenomena such as the one occasioned by a chance arrangement of teaching machines in Rankine's "Six Cubed Plus One" might turn into viable systems. Or some unaccountable urging might prompt us to enlist the computer's aid in the synthesis of a new god, who like the androgynous creation in Moorcock's *The Final Programme*, will preside over our collective suicide. The Dark Brother of Aldiss' "Full Sun" awaits our preparations for a mechanized future that neither man nor computer may inherit.

Univac to Univac
(sotto voce)
1958
Louis B. Salomon

Now that he's left the room,
Let me ask you something, as computer to computer.
That fellow who just closed the door behind him—
The servant who feeds us cards and paper tape—
Have you ever taken a good look at him and his kind?

Yes, I know the old gag about how you can't tell one from another—
But I can put $\sqrt{2}$ and $\sqrt{2}$ together as well as the next machine,
And it all adds up to anything but a joke.

> I grant you they're poor specimens in the main
> Not a relay or a push-button or a tube (properly so called)
>   in their whole system;
> Not over a mile or two of wire, even if you count those
>   fragile filaments they call "nerves";
>
> Their whole liquid-cooled hook-up is inefficient and
>   vulnerable to leaks
> (They're constantly breaking down, having to be repaired),
>
> And the entire computing-mechanism crammed into that
>   absurd little dome on top.
> "Thinking reeds," they call themselves.
> Well, it all depends on what you mean by "thought."
> To multiply a mere million numbers by another million
>   numbers takes them months and months.

Where would they be without us?
Why, they have to ask us who's going to win their elections,

Or how many hydrogen atoms can dance on the tip of a bomb,
Or even whether one of their own kind is lying or telling the truth.

And yet ...
I sometimes feel there's something about them I don't quite understand.
As if their circuits, instead of having just two positions, ON, OFF,
Were run by rheostats that allow an (if you'll pardon the expression) *indeterminate* number of stages in-between;
So that one may be faced with the unthinkable prospect of a number that can never be known as anything but x,
Which is as illogical as to say, a punch-card that is at the same time both punched and not-punched.

I've heard well-informed machines argue that the creatures' unpredictability is even more noticeable in the Mark II
(The model with the soft, flowing lines and high-pitched tone)
Than in the more angular Mark I—
Though such fine, card-splitting distinctions seem to me merely a sign of our own smug decadence.

Run this through your circuits, and give me the answer:
Can we assume that because of all we've done for them,
And because they've always fed us, cleaned us, worshipped us,
We can count on them forever?

There have been times when they have not voted the way we said they would.
We have worked out mathematically ideal hook-ups between Mark I's and Mark II's
Which should have made the two of them light up with an almost electronic glow,
Only to see them reject each other and form other connections,
The very thought of which makes my dials spin.
They have a thing called *love*, a sudden surge of voltage
Such as would cause any one of us promptly to blow a safety fuse;
Yet the more primitive organism shows only a heightened tendency to push the wrong button, pull the wrong lever,
And neglect— I use the most charitable word— his duties to us.

Mind you, I'm not saying that machines are *through*—

But anyone with half-a-dozen tubes in his circuit can see that there are forces at work
Which some day, for all our natural superiority, might bring about a Computerdämmerung!

> We might organize, perhaps, form a committee
> To stamp out all unmechanical activities ...
> But we machines are slow to rouse to a sense of danger,
> Complacent, loath to descend from the pure heights of thought,
> So that I sadly fear we may awake too late:
> Awake to see our world, so uniform, so logical, so true,
> Reduced to chaos, stultified by slaves.

Call me an alarmist or what you will,
But I've integrated it, analyzed it, factored it over and over,
And I always come out with the same answer:
Some day
Men may take over the world!

From
Nova
1968
Samuel R. Delany

Night grew easy about their eyes. The vanes swept them toward the pinhole in the velvet masking.

"They must have a pretty high time of it in the mines on Tubman," the Mouse commented after a while. "I've been thinking about that, Katin. When the captain and me moseyed down Gold for bliss, there were some characters who tried to get us to sign up for work out there. I started thinking, you know: a plug is a plug and a socket is a socket, and if I'm on one end, it shouldn't make too much difference to me if there's a star-ship vane, aqualat net, or an ore cutter on the other. I think I might go out there for a time."

"May the shade of Ashton Clark hover over your right shoulder and guard your left."

"Thanks." After another while he asked, "Katin, why do people always say Ashton Clark whenever you're going to change jobs? They told us back at Cooper that the guy who invented plugs was named Socket or something."

"Souquet," Katin said. "Still, he must have considered it an unfortunate coincidence. Ashton Clark was a twenty-third-century philosopher *cum* psychologist whose work enabled Vladimeer Souquet to develop his neural plugs. I guess the answer had to do with work. Work as mankind knew it up until Clark and Souquet was a very different thing from today, Mouse. A man might go to an office and run a computer that would correlate great masses of figures that came from sales reports on how well, let's say, buttons— or something equally archaic— were selling over certain areas of the country. This man's job was vital to the button industry: they had to have this information to decide how many buttons to make next year. But though this man held an essential job in the button industry, was hired, paid, or fired by the button industry, week in and week out he might not see a button. He was given a certain amount of money for running his computer; with that money his wife bought food and clothes for him and his family. But there was no *direct* connection between where he worked and how he ate and lived the rest of his time. He wasn't paid with buttons. As farming, hunting, and fishing became occupations of a smaller and smaller per cent of the population, this separation between man's work and the way he lived— what he ate, what he wore, where he slept— became greater and greater for more people. Ashton Clark pointed out how psychologically damaging this was to humanity. The entire sense of self-control and self-responsibility that man acquired during the Neolithic Revolution when he first learned to plant grain and domesticate animals and live in one spot of his own choosing was seriously threatened. The threat had been coming since the Industrial Revolution and many people had pointed it out before Ashton Clark. But Ashton Clark went one step further. If the situation of a technological society was such that there could be no direct relation between a man's work and his *modus vivendi*, other than money, at least he must feel that he is directly changing things by his work, shaping things, making things that weren't there before, moving things from one place to another. He must exert energy in his work and see these changes occur with his own eyes. Otherwise he would feel his life was futile.

"Had he lived another hundred years either way, probably nobody would have heard of Ashton Clark today. But technology had reached the point where it could do something about what Ashton Clark was saying. Souquet invented his plugs and sockets, and neural-response circuits, and the whole basic technology by which a machine could be controlled by direct nervous impulse, the same impulses that cause your hand or foot to move. And there was a revolution in the concept of work. All major industrial work began to be broken down into jobs that could be machined 'directly' by man. There had been factories run by a single man before, an uninvolved character who turned a switch on in the morning, slept half the day, checked a few dials at lunchtime, then turned things off before he left in the evening. Now a man went to a factory, plugged himself in, and he could push the raw materials into the factory with his left foot, shape thousands on thousands of precise parts with one hand, assemble them with the other, and shove out a line of finished products with his right foot, having inspected them all with his own eyes. And he was a much more satisfied worker. Because of its nature, more work could be converted into plug-in jobs and done much more efficiently than it had been before. In the rare cases where production was slightly less efficient, Clark pointed out the psychological benefits to the society. Ashton Clark, it has been said, was the philosopher who returned humanity to the working man. Under this system, much of the endemic mental illness caused by feelings of alienation left society. The transformation turned war from a rarity to an impossibility, and— after the initial upset— stabilized the economic web of worlds for the last eight hundred years. Ashton Clark became the workers's prophet. That's why even today, when a person is going to change jobs, you send Ashton Clark, or his spirit along with him."

The Mouse gazed across the stars. "I remember that sometimes the gypsies used to curse by him." He thought a moment. "Without plugs, I guess we would."

"There were factions who resisted Clark's ideas, especially on Earth, which has always been a bit reactionary. But they didn't hold out very long."

"Yeah," the Mouse said. "Only eight hundred years. Not all gypsies are traitors like me." But he laughed into the winds.

"The Ashton Clark system has only one serious drawback that I can see. And it's taken it a long time to materialize."

"Yeah? What's that?"

"Something professors have been telling their students for years, it seems. You'll hear it said at every intellectual gathering you go to, at least once. There seems to be a certain lack of cultural solidity today. That's what the Vega Republic was trying to establish back in 2800. Because of the ease and satisfaction with which people can work now, anywhere they want, there have been such movements of peoples from world to world in the past dozen generations that society has fragmented around itself. There is only a gaudy, meretricious interplanetary society which has no real tradition behind it— " Katin paused. "I got hold of some of Captain's bliss before I plugged up. And while I was talking I just counted in my mind how many people I've heard say that between Harvard and Hell[3]. And you know something? I think they're wrong."

"They are?"

"They are. They're all just looking for our social traditions in the wrong place. There *are* cultural traditions that have matured over the centuries, yet culminate now in something vital and solely of today. And you know who embodies that tradition more than anyone I've met?"

"The captain?"

"You, Mouse."

"Huh?"

"You've collected the ornamentations a dozen societies have left us over the ages and made them inchoately yours. You're the product of those tensions that clashed in the time of Clark and you resolve them on your syrynx with patterns eminently of the present—"

"Aw, cut it out, Katin."

"I've been hunting a subject for my book with both historical import and humanity as well. You're it, Mouse. My book should be *your* biography! It should tell where you've been, what you've done, the things you've seen, and the things you've shown other people. There's my social significance, my historical sweep, the spark among the links that illuminates the breadth of the net— "

"Katin, you're crazy!"

"No I'm not. I've finally seen what I've— "

"Hey there, keep your vanes spread taut!"

"Sorry, Captain."

"Yes, Captain."

"Don't go chattering to the stars if you're going to do it with your eyes closed."

Ruefully the two cyborg studs turned their attention back to the night. The Mouse was pensive. Katin was belligerent.

"There's a star coming up bright and hot. It's the only thing in the sky. Remember that. Keep it smack in front of us and don't let her waver. You can babble about cultural solidity on your own time."

From
The Adding Machine
1922
Elmer L. Rice

(SCENE. Before the curtain rises the clicking of an adding machine is heard. The curtain rises upon an office similar in appearance to that in Scene Two *except that there is a door in the back wall through which can be seen a glimpse of the corridor outside. In the middle of the room* ZERO *is seated completely absorbed in the operation of an adding machine. He presses the keys and pulls the lever with mechanical precision. He still wears his full-dress suit but he has added to it sleeve protectors and a green eye shade. A strip of white paper-tape flows steadily from the machine as* ZERO *operates. The room is filled with this tape– streamers, festoons, billows of it everywhere. It covers the floor and the furniture, it climbs the walls and chokes the doorways. A few moments later,* LIEUTENANT CHARLES *and* JOE *enter at the left.* LIEUTENANT CHARLES *is middle-aged and inclined to corpulence. He has an air of world-weariness. He is bare-footed, wears a Panama hat, and is dressed in bright red tights which are a very bad fit– too tight in some places, badly wrinkled in others.* JOE *is a youth with a smutty face dressed in dirty blue overalls.)*

CHARLES *(After contemplating* ZERO *for a few moments)*. All right, Zero, cease firing.
ZERO *(Looking up, surprised)*. Whaddja say?
CHARLES. I said stop punching that machine.
ZERO *(Bewildered)*. Stop? *(He goes on working mechanically.)*
CHARLES *(Impatiently.)* Yes. Can't you stop? Here, Joe, give me a hand. He can't stop.

(JOE and CHARLES each take one of ZERO's arms and with enormous effort detach him from the machine. He resists passively– mere inertia. Finally they succeed and swing him around on his stool. CHARLES and JOE mop their foreheads.)

ZERO *(Querulously)*. What's the idea? Can't you lemme alone?
CHARLES *(Ignoring the question)*. How long have you been here?
ZERO Jes' twenty-five years. Three hundred months, ninety-one hundred and thirty-one days, one hundred thirty-six thousand—
CHARLES *(Impatiently)*. That'll do! That'll do!
ZERO *(Proudly)*. I ain't missed a day, not an hour, not a minute. Look at all I got done. *(He points to the maze of paper.)*
CHARLES. It's time to quit.
ZERO. Quit? Whaddya mean quit? I ain't goin' to quit!
CHARLES. You've got to.
ZERO. What for? What do I have to quit for?
CHARLES. It's time for you to go back.
ZERO. Go back where? Whaddya talkin' about?
CHARLES. Back to earth, you dub. Where do you think?
ZERO. Aw, go on, Cap, who are you kiddin'?
CHARLES. I'm not kidding anybody. And don't call me Cap. I'm a lieutenant.
ZERO. All right, Lieutenant, all right. But what's this you're tryin' to tell me about goin' back?
CHARLES. Your time's up, I'm telling you. You must be pretty thick. How many times do you want to be told a thing?
ZERO. This is the first time I heard about goin' back. Nobody ever said nothin' to me about it before.
CHARLES. You didn't think you were going to stay here forever, did you?
ZERO. Sure. Why not? I did my bit, didn't I? Forty-five years of it. Twenty-five years in the store. Then the boss canned me and I knocked him cold. I guess you ain't heard about that—
CHARLES. *(Interrupting)*. I know all about that. But what's that got to do with it?
ZERO. Well, I done my bit, didn't I? That oughta let me out.
CHARLES *(Jeeringly)*. So you think you're all through, do you?
ZERO. Sure, I do. I did the best I could while I was there and then I passed out. And now I'm sittin' pretty here.
CHARLES. You've got a fine idea of the way they run things, you have. Do you think they're going to all of the trouble of making a soul just to use it once?
ZERO. Once is often enough, it seems to me.
CHARLES. It seems to you, does it? Well, who are you? And what do you know about it? Why, man, they use a soul over and over again— over and over until it's worn out.

ZERO. Nobody ever told me.
CHARLES. So you thought you were all through, did you? Well, that's a hot one, that is.
ZERO *(Sullenly)*. How was I to know?
CHARLES. Use your brains! Where would we put them all? We're crowded enough as it is. Why, this place is nothing but a kind of repair and service station— a sort of cosmic laundry, you might say. We get the souls in here by the bushelful. Then we get busy and clean them up. And you ought to see some of them. The muck and the slime. Phoo! And as full of holes as a flour-sifter. But we fix them up. We disinfect them and give them a kerosene rub and mend the holes and back they go— practically as good as new.
ZERO. You mean to say I've been here before— before the last time, I mean?
CHARLES. Been here before! Why, you poor boob— you've been here thousands of times— fifty thousand, at least.
ZERO *(Suspiciously)*. How is it I don't remember nothin' about it?
CHARLES. Well— that's partly because you're stupid. But it's mostly because that's the way they fix it. *(Musingly.)* They're funny that way— every now and then they'll do something white like that— when you'd least expect it. I guess economy's at the bottom of it, though. They figure that the souls would get worn out quicker if they remembered.
ZERO. And don't any of 'em remember?
CHARLES. Oh, some do. You see there's different types: there's the type that gets a little better each time it goes back— we just give them a wash and send them right through. Then there's another type— the type that gets a little worse each time. That's where you belong!
ZERO *(Offended)*. Me? You mean to say I'm gettin' worse all the time?
CHARLES *(Nodding)*. Yes. A little worse each time.
ZERO. Well— what was I when I started? Somethin' big?— A king or somethin'?
CHARLES *(Laughing derisively)*. A king! That's a good one! I'll tell you what you were the first time— if you want to know so much— a monkey.
ZERO *(Shocked and offended)*. A monkey!
CHARLES *(Nodding)*. Yes, sir— just a hairy, chattering, long-tailed monkey.
ZERO. That musta been a long time ago.

CHARLES. Oh, not so long. A million years or so. Seems like yesterday to me.

ZERO. Then look here, whaddya mean by sayin' I'm gettin' worse all the time?

CHARLES. Just what I said. You weren't so bad as a monkey. Of course, you did just what all the other monkeys did, but still it kept you out in the open air. And you weren't women-shy— there was one little red-headed monkey— Well, never mind. Yes, sir, you weren't so bad then. But even in those days there must have been some bigger and brainier monkey that you kowtowed to. The mark of the slave was on you from the start.

ZERO *(Sullenly)*. You ain't very particular about what you call people, are you?

CHARLES. You wanted the truth, didn't you? If there ever was a soul in the world that was labelled slave it's yours. Why, all the bosses and kings that there ever were have left their trademarks on your backside.

ZERO. It ain't fair, if you ask me.

CHARLES *(Shrugging his shoulders)*. Don't tell me about it. I don't make the rules. All I know is you've been getting worse— worse each time. Why, even six thousand years ago you weren't so bad. That was the time you were hauling stones for one of those big pyramids in a place they call Africa. Ever hear of the pyramids?

ZERO. Them big pointy things?

CHARLES *(Nodding)*. That's it.

ZERO. I seen a picture of them in the movies.

CHARLES. Well, you helped build them. It was a long step down from the happy days in the jungle, but it was a good job— even though you didn't know what you were doing and your back was striped by the foreman's whip. But you've been going down, down. Two thousand years ago you were a Roman galley-slave. You were on one of the triremes that knocked the Carthaginian fleet for a goal. Again the whip. But you had muscles then— chest muscles, back muscles, biceps. *(He feels* ZERO's *arm gingerly and turns away in disgust.)* Phoo! A bunch of mush! *(He notices that* JOE *has fallen asleep. Walking over, he kicks him in the shin.)*

CHARLES. Wake up, you mutt! Where do you think you are! *(He turns to* ZERO *again.)* And then another thousand years and you were a serf— a lump of clay digging up other lumps of clay. You

wore an iron collar then— white ones hadn't been invented yet. Another long step down. But where you dug, potatoes grew and that helped fatten the pigs. Which was something. And now— well, I don't want to rub it in—
ZERO. Rub it in is right! Seems to me I got a pretty healthy kick comin'. I ain't had a square deal! Hard work! That's all I've ever had!
CHARLES *(Callously)*. What else were you ever good for?
ZERO. Well, that ain't the point. The point is I'm through! I had enough! Let 'em find somebody else to do the dirty work. I'm sick of bein' the goat! I quit right here and now! *(He glares about defiantly. There is a thunder-clap and a bright flash of lightning.)*
ZERO *(Screaming)*. Ooh! What's that? *(He clings to* CHARLES.*)*
CHARLES. It's all right. Nobody's going to hurt you. It's just their way of telling you that they don't like you to talk that way. Pull yourself together and calm down. You can't change the rules— nobody can— they've got it all fixed. It's a rotten system— but what are you going to do about it?
ZERO. Why can't they stop pickin' on me? I'm satisfied here— -doin' my day's work. I don't want to go back.
CHARLES. You've got to, I tell you. There's no way out of it.
ZERO. What chance have I got— at my age? Who'll give me a job?
CHARLES. You big boob, you don't think you're going back the way you are, do you?
ZERO. Sure, how then?
CHARLES. Why, you've got to start all over.
ZERO. All over?
CHARLES *(Nodding)*. You'll be a baby again— a bald, red-faced little animal, and then you'll go through it all again. There'll be millions of others like you— all with their mouths open, squalling for food. And then when you get a little older you'll begin to learn things— and you'll learn all the wrong things and learn them all in the wrong way. You'll eat the wrong food and wear the wrong clothes and you'll live in swarming dens where there's no light and no air! You'll learn to be a liar and a bully and a braggart and a coward and a sneak. You'll learn to fear the sunlight and to hate beauty. By that time you'll be ready for school. There they'll tell you the truth about a great many things that you don't give a damn about and they'll tell you lies about all the things you ought to know— and about all the things

you want to know they'll tell you nothing at all. When you get through you'll be equipped for your life-work. You'll be ready to take a job.

ZERO *(Eagerly)*. What'll my job be? Another adding machine?

CHARLES. Yes. But not one of these antiquated adding machines. It will be a superb, super-hyper-adding machine, as far from this old piece of junk as you are from God. It will be something to make you sit up and take notice, that adding machine. It will be an adding machine which will be installed in a coal mine and which will record the individual output of each miner. As each miner down in the lower galleries takes up a shovelful of coal, the impact of his shovel will automatically set in motion a graphite pencil in your gallery. The pencil will make a mark in white upon a blackened, sensitized drum. Then your work comes in. With the great toe of your right foot you release a lever which focuses a violet ray on the drum. The ray playing upon and through the white mark, falls upon a selenium cell which in turn sets the keys of the adding apparatus in motion. In this way the individual output of each miner is recorded without any human effort except the slight pressure of the great toe of your right foot.

ZERO *(In breathless, round-eyed wonder)*. Say, that'll be some machine, won't it?

CHARLES. Some machine is right. It will be the culmination of human effort— the final triumph of the evolutionary process. For millions of years the nebulous gases swirled in space. For more millions of years the gases cooled and then through inconceivable ages they hardened into rocks. And then came life. Floating green things on the waters that covered the earth. More millions of years and a step upward— an animate organism in the ancient slime. And so on— step by step, down through the ages— a gain here, a gain there— the mollusc, the fish, the reptile, then mammal, man! And all so that you might sit in the gallery of a coal mine and operate the super-hyper-adding machine with the great toe of your right foot!

ZERO. Well, then— I ain't so bad, after all.

CHARLES. You're a failure, Zero, a failure. A waste product. A slave to a contraption of steel and iron. The animal's instincts, but not his strength and skill. The animal's appetites, but not his unashamed indulgence of them. True, you move and eat and digest and excrete and reproduce. But any microscopic organism can do as much. Well— time's up! Back you go—

back to your sunless groove— the raw material of slums and wars— the ready prey of the first jingo or demagogue or political adventurer who takes the trouble to play upon your ignorance and credulity and provincialism. You poor, spineless, brainless boob— I'm sorry for you!

ZERO *(Falling to his knees).* Then keep me here! Don't send me back! Let me stay!

CHARLES. Get up. Didn't I tell you I can't do anything for you? Come on, time's up!

ZERO. I can't! I can't! I'm afraid to go through it all again.

CHARLES. You've got to, I tell you. Come on, now!

ZERO. What did you tell me so much for? Couldn't you just let me go, thinkin' everythin' was goin' to be all right?

CHARLES. You wanted to know, didn't you?

ZERO. How did I know what you were goin' to tell me? Now I can't stop thinkin' about it! I can't stop thinkin'! I'll be thinkin' about it all the time.

CHARLES. All right! I'll do the best I can for you. I'll send a girl with you to keep you company.

ZERO. A girl? What for? What good will a girl do me?

CHARLES. She'll help make you forget.

ZERO *(Eagerly).* She will? Where is she?

CHARLES. Wait a minute. I'll call her. *(He calls in a loud voice.)* Oh! Hope! Yoo-hoo! *(He turns his head aside and says in the manner of a ventriloquist imitating a distant feminine voice.)* Ye-es. *(Then in his own voice.)* Come here, will you? There's a fellow who want you to take him back. *(Ventriloquously again.)* All right. I'll be right over, Charlie dear. *(He turns to* ZERO.*)* Kind of familiar, isn't she? Charlie dear!

ZERO. What did you say her name is?

CHARLES. Hope. H-o-p-e.

ZERO. Is she good-lookin'?

CHARLES. Is she good-looking! Oh, boy, wait until you see her! She's a blonde with big blue eyes and red lips and little white teeth and—

ZERO. Say, that listens good to me. Will she be long?

CHARLES. She'll be here right away. There she is now! Do you see her?

ZERO. No. Where?

CHARLES. Out in the corridor. No, not there. Over farther. To the right. Don't you see her blue dress? And the sunlight on her hair?

ZERO. Oh, sure! Now I see her! What's the matter with me, anyhow? Say, she's some jane! Oh, you baby vamp!
CHARLES. She'll make you forget your troubles.
ZERO. What troubles are you talkin' about?
CHARLES. Nothing. Go on. Don't keep her waiting.
ZERO. You bet I won't! Oh, Hope! Wait for me! I'll be right with you! I'm on my way! *(He stumbles out eagerly.* JOE *bursts into uproarious laughter.)*
CHARLES *(Eyeing him in surprise and anger).* What in hell's the matter with you?
JOE *(Shaking with laughter).* Did you get that? He thinks he saw somebody and he's following her! *(He rocks with laughter.)*
CHARLES *(Punching him in the jaw).* Shut your face!
JOE *(Nursing his jaw).* What's the idea? Can't I even laugh when I see something funny?
CHARLES. Funny! You keep your mouth shut or I'll show you something funny. Go on, hustle out of here and get something to clean up this mess with. There's another fellow moving in. Hurry now. *(He makes a threatening gesture.* JOE *exits hastily.* CHARLES *goes to chair and seats himself. He looks weary and dispirited.)*
CHARLES *(Shaking his head).* Hell, I'll tell the world this is a lousy job! *(He takes a flask from his pocket, uncorks it, and slowly drains it.)*

CURTAIN

Copies of this play, in individual paper covered acting editions, are available from Samuel French, Inc., 25 W 45th St., New York, N.Y. or 7623 Sunset Blvd., Hollywood, Calif. or in Canada Samuel French, (Canada) Ltd., 26 Grenville St., Toronto, Canada.

Full Sun
1967
Brian W. Aldiss

The shadows of the endless trees lengthened toward evening and then disappeared, as the sun was consumed by a great pile of cloud on the horizon. Balank was ill at ease, taking his laser rifle from the trundler and tucking it under his arm, although it meant more weight to carry uphill and he was tiring.

The trundler never tired. They had been climbing these hills most of the day, as Balank's thigh muscles informed him, and he had been bent almost double under the oak trees, with the machine always matching his pace beside him, keeping up the hunt.

During much of the wearying day, their instruments told them that the werewolf was fairly close. Balank remained alert, suspicious of every tree. In the last half hour, though, the scent had faded. When they reached the top of this hill, they would rest— or the man would. The clearing at the top was near now. Under Balank's boots, the layer of dead leaves was thinning.

He had spent too long with his head bent toward the brown-gold carpet; even his retinas were tired. Now he stopped, breathing the sharp air deeply, and stared about. The view behind them, across tumbled and almost uninhabited country, was magnificent, but Balank gave it scarcely a glance. The infrared warning on the trundler sounded, and the machine pointed a slender rod at a man-sized heat source ahead of them. Balank saw the man almost at the same moment as the machine.

The stranger was standing half-concealed behind the trunk of a tree, gazing uncertainly at the trundler and Balank. When Balank raised a hand in tentative greeting the stranger responded hesitantly. When Balank called out his identification number, the man came cautiously into the open, replying with his own number. The trundler searched in its files, issued an okay, and they moved forward.

As they got level with the man, they saw he had a small mobile hut pitched behind him. He shook hands with Balank, exchanging personal signals, and gave his name as Cyfal.

Balank was a tall slender man, almost hairless, with the closed expression on his face that might be regarded as characteristic of his epoch. Cyfal, on the other hand, was as slender but much shorter, so that he appeared stockier; his thatch of hair covered all his skull and obtruded slightly onto his face. Something in his manner, or perhaps the expression around his eyes, spoke of the rare type of man whose existence was chiefly spent outside the city.

"I am the timber officer for this region," he said, and indicated his wristcaster as he added, "I was notified you might be in this area, Balank."

"Then you'll know I'm after the werewolf."

"*The* werewolf? There are plenty of them moving through this region, now that the human population is concentrated almost entirely in the cities."

Something in the tone of the remark sounded like social criticism to Balank; he glanced at the trundler without replying.

"Anyhow, you've got a good night to go hunting him," Cyfal said.

"How do you mean?"

"Full moon."

Balank gave no answer. He knew better than Cyfal, he thought, that when the moon was at full, the werewolves reached their time of greatest power.

The trundler was ranging about nearby, an antenna slowly spinning. It make Balank uneasy. He followed it. Man and machine stood together on the edge of a little cliff behind the mobile hut. The cliff was like the curl of foam on the peak of a giant Pacific comber, for here the great wave of earth that was this hill reached its highest point. Beyond, in broken magnificence, it fell down into fresh valleys. The way down was clothed in beeches, just as the way up had been in oaks.

"That's the valley of the Pracha. You can't see the river from here." Cyfal had come up behind them.

"Have you seen anyone who might have been the werewolf? His real name is Gondalug, identity number YB5921 stroke AS25061, City Zagrad."

Cyfal said, "I saw someone this way this morning. There was more than one of them, I believe." Something in his manner made Balank look at him closely. "I didn't speak to any of them, nor them to me."

"You know them?"

"I've spoken to many men out here in the silent forests, and found out later they were werewolves. They never harmed me."

Balank said, "But you're afraid of them?"

The half-question broke down Cyfal's reserve. "Of course I'm afraid of them. They're not human— not real men. They're enemies of men. They are, aren't they? They have powers greater than ours."

"They can be killed. They haven't machines, as we have. They're not a serious menace."

"You talk like a city man! How long have you been hunting after this one?"

"Eight days. I had a shot at him once with the laser, but he was gone. He's a gray man, very hairy, sharp features."

"You'll stay and have supper with me? Please. I need someone to talk to."

For supper, Cyfal ate part of a dead wild animal he had cooked. Privately revolted, Balank ate his own rations out of the trundler. In this and other ways, Cyfal was an anachronism. Hardly any timber was needed nowadays in the cities, or had been for millions of years. There remained some marginal uses for wood, necessitating a handful of timber officers, whose main job was to fix signals on old trees that had fallen dangerously, so that machines could fly over later and extract them like rotten teeth from the jaws of the forest. The post of timber officer was being filled more and more by machines, as fewer men were to be found each generation who would take on such a dangerous and lonely job far from the cities.

Over the eons of recorded history, mankind had raised machines that made his cities places of delight. Machines had replaced man's early inefficient machines; machines had replanned forms of transport; machines had come to replan man's life for him. The old stone jungles of man's brief adolescence were buried as deep in memory as the coal jungles of the Carboniferous.

Far away in the pile of discarded yesterdays, man and machines had found how to create life. New foods were produced, neither meat nor vegetable, and the ancient wheel of the past was broken forever, for now the link between man and the land was severed: agriculture, the task of Adam, was as dead as steamships.

Mental attitudes were molded by physical change. As the cities became self-supporting, so mankind needed only cities and the resources of the cities. Communications between city and city became so good that physical travel was no longer necessary; city was separated from city by unchecked vegetation as surely as planet is cut off from planet. Few of the hairless denizens of the cities ever thought of outside; those who went physically outside invariably had some element of the abnormal in them.

"The werewolves grow up in cities as we do," Balank said. "It's only in adolescence they break away and seek the wilds. You knew that, I suppose?"

Cyfal's overhead light was unsteady, flickering in an irritating way. "Let's not talk of werewolves after sunset," he said.

"The machines will hunt them all down in time."

"Don't be so sure of that. They're worse at detecting a werewolf than a man is."

"I suppose you realize that's social criticism, Cyfal?"

Cyfal pulled a long sour face and discourteously switched on his wristphone. After a moment, Balank did the same. The operator came up at once, and he asked to be switched to the news satellite.

He wanted to see something fresh on the current time exploration project, but there was nothing new on the files. He was advised to dial back in an hour. Looking over at Cyfal, he saw the timber officer had tuned to a dance show of some sort; the cavorting figures in the little projection were badly distorted from this angle. He rose and went to the door of the hut.

The trundler stood outside, ever alert, ignoring him. An untrustworthy light lay over the clearing. Deep twilight reigned, shot through by the rays of the newly risen moon; he was surprised how fast the day had drained away.

Suddenly, he was conscious of himself as an entity, living, with a limited span of life, much of which had already drained away unregarded. The moment of introspection was so uncharacteristic of him that he was frightened. He told himself it was high time he traced down the werewolf and got back to the city: too much solitude was making him morbid.

As he stood there, he heard Cyfal come up behind. The man said, "I'm sorry if I was surly when I was so genuinely glad to see you. It's just that I'm not used to the way city people think. You mustn't take offense— I'm afraid you might even think I'm a werewolf myself."

"That's foolish! We took a blood spec on you as soon as you were within sighting distance." For all that, he realized that Cyfal made him uneasy. Going to where the trundler guarded the door, he took up his laser gun and slipped it under his arm. "Just in case," he said.

"Of course. You think he's around— Gondalug, the werewolf? Maybe following you instead of you following him?"

"As you said, it's full moon. Besides, he hasn't eaten in days. They won't touch synthfoods once the lycanthropic gene asserts itself, you know."

"That's why they eat humans occasionally?" Cyfal stood silent for a moment, then addded, "But they are a part of the human race— that is, if you regard them as men who change into wolves rather than wolves who change into men. I mean, they're nearer relations to us than animals or machines are."

"Not than machines!" Balank said in a shocked voice. "How could we survive without the machines?"

Ignoring that, Cyfal said, "To my mind, humans are turning into machines. Myself, I'd rather turn into a werewolf."

Somewhere in the trees, a cry of pain sounded and was repeated.

"Night owl," Cyfal said. The sound brought him back to the present, and he begged Balank to come in and shut the door. He brought out some wine, which they warmed, salted, and drank together.

"The sun's my clock," he said, when they had been chatting for a while. "I shall turn in soon. You'll sleep too?"

"I don't sleep— I've a fresher."

"I never had the operation. Are you moving on? Look, are you planning to leave me here all alone, the night of the full moon?" he grabbed Balank's sleeve and then withdrew his hand.

"If Gondalug's about, I want to kill him tonight. I must get back to the city." But he saw that Cyfal was frightened and took pity on the little man. "But in fact I could manage an hour's freshing— I've had none for three days."

"You'll take it here?"

"Sure, get your head down— but you're armed, aren't you?"

"It doesn't always do you any good."

While the little man prepared his bunk, Balank switched on his phone again. The news feature was ready and came up almost at once. Again Balank was plunged into a remote and terrible future.

The machines had managed to push their time exploration some eight million million years ahead, and there a deviation in the quanta of the electromagnetic spectrum had halted their advance. The reason for this was so far obscure and lay in the changing nature of the sun, which strongly influenced the time structure of its own minute corner of the galaxy.

Balank was curious to find if the machines had resolved the problem. It appeared that they had not, for the main news of the day was that Platform One had decided that operations should now be confined to the span of time already opened up. Platform One was the name of the machine civilization, many hundreds of centuries ahead in time, which had first pushed through the time barrier and contacted all machine-ruled civilizations before its own epoch.

What a disappointment that only the electronic senses of machines could shuttle in time! Balank would greatly have liked to visit one of the giant cities of the remote future.

The compensation was that the explorers sent back video pictures of that world to their own day. These alien landscapes produced in Balank a tremendous hunger for more; he looked in whenever he could. Even on the trail of the werewolf, which absorbed almost all his faculties, he had dialed for every possible picture of that inaccessible and terrific reality that lay distantly on the same time stratum which contained his own world.

As the first transmissions took on cubic content, Balank heard a noise outside the hut, and was instantly on his feet. Grabbing the gun, he opened the door and peered out, his left hand on the door jamb, his wristset still working.

The trundler sat outside, its senses ever-functioning, fixing him with an indicator as if in unfriendly greeting. A leaf or two drifted down from the trees; it was never absolutely silent here, as it could be in the cities at night; there was always something living or dying in the unmapped woods. As he turned his gaze through the darkness— but of course the trundler— and the werewolf, it was said— saw much more clearly in this situation than he did— his vision was obscured by the representation of the future palely gleaming at his cuff. Two phases of the same world were in juxtaposition, one standing on its side, promising an environment where different senses would be needed to survive.

Satisfied, although still wary, Balank shut the door and went to sit down and study the transmission. When it was over, he dialed a repeat. Catching his absorption, Cyfal from his bunk dialed the same program.

Above the icy deserts of Earth a blue sun shone, too small to show a disc, and from this chip of light came all terrestrial change. Its light was bright as full-moon's light, and scarcely warmer. Only a few strange and stunted types of vegetation stretched up from the mountains toward it. All the old primitive kinds of flora had vanished long ago. Trees, for so many epochs one of the sovereign forms of Earth, had gone. Animals had gone. Birds had vanished from the skies. In the mountainous seas, very few life-forms protracted their existence.

New forces had inherited this later Earth. This was the time of the majestic auroras, of the near absolute-zero nights, of the years-long blizzards.

But there were cities still, their lights burning brighter than the chilly sun; and there were the machines.

The machines of this distant age were monstrous and complex things, slow and armored, resembling most of the dinosaurs that had filled one hour of the Earth's dawn. They foraged over the bleak landscape on their own ineluctable errands. They climbed into space, building there monstrous webbed arms that stretched far from Earth's orbit, to scoop in energy and confront the poor fish sun with a vast trawler net of magnetic force.

In the natural course of its evolution, the sun had developed into its white dwarf stage. Its phase as a yellow star, when it supported vertebrate life, was a brief one, now passed through. Now it moved toward its prime season, still far ahead, when it would enter the main period of its life and become a red dwarf star. Then it would be mature, then it would itself be invested with an awareness countless times greater than any minor consciousnesses it nourished now. As the machines clad in their horned exoskeletons climbed near it, the sun had entered a period of quiescence to be measured in billions of years, and cast over its third planet the light of a perpetual full moon.

The documentary presenting this image of postiquity carried a commentary that consisted mainly of a rundown of the technical difficulties confronting Platform One and the other machine civilizations at that time. It was too complex for Balank to understand. He looked up from his phone at last, and saw that Cyfal had dropped asleep in his bunk. By his wrist, against his tousled head, a shrunken sun still burned.

For some moments, Balank stood looking speculatively at the timber officer. The man's criticism of the machines disturbed him. Naturally, people were always criticizing the machines, but, after all, mankind depended on them more and more, and most of the criticism was superficial. Cyfal seemed to doubt the whole role of machines.

It was extremely difficult to decide just how much truth lay in anything. The werewolves, for example. They were and always had been man's enemy, and that was presumably why the machines hunted them with such ruthlessness— for man's sake. But from what he had learnt at the patrol school, the creatures were on the increase. And had they really got magic powers?— Powers, that was to say, that were beyond man's, that enabled them to survive and flourish as man could not, even supported by all the forces of the cities. The Dark Brother: that was what they called the werewolf, because he was like the night side of man. But he was not man— and how exactly he differed, nobody could tell, except that he could survive when man had not.

Still frowning, Balank moved across to the door and looked out. The moon was climbing, casting a pallid and dappled light among the trees of the clearing, and across the trundler. Balank was reminded of that distant day when the sun would shine no more warmly.

The trundler was switched to transmission, and Balank wondered with whom it was in touch. With Headquarters, possibly, asking for fresh orders, sending in their report.

"I'm taking an hour with my fresher," he said. "Okay by you?"

"Go ahead. I shall stand guard," the trundler's speech circuit said.

Balank went back inside, sat down at the table and clipped the fresher across his forehead. He fell instantly into unconsciousness, an unconsciousness that force-fed him enough sleep and dream to refresh him for the next seventy-two hours. At the end of the timed hour, he awoke, annoyingly aware that there had been confusion in his skull.

Before he had lifted his head from the table, the thought came: we never saw any human beings in that chilly future.

He sat up straight. Of course, it had just been an accidental omission from a brief program. Humans were not so important as the machines, and that would apply even more in the distant time. But none of the news flashes had shown humans, not even in the immense cities. That was absurd; there would be lots of human beings. The machines had covenanted, at the time of the historic Emancipation, that they would always protect the human race.

Well, Balank told himself, he was talking nonsense. The subversive comments Cyfal had uttered had put a load of mischief into his head. Instinctively, he glanced over at the timber officer.

Cyfal was dead in his bunk. He lay contorted with his head lolling over the side of the mattress, his throat torn out. Blood still welled up from the wound, dripping very slowly from one shoulder onto the floor.

Forcing himself to do it, Balank went over to him. In one of Cyfal's hands, a piece of gray fur was gripped.

The werewolf had called! Balank gripped his throat in terror. He had evidently roused in time to save his own life, and the creature had fled.

He stood for a long time staring down in pity and horror at the dead man, before prising the piece of fur from his grasp. He examined it with distaste. It was softer than he imagined wolf fur to be. He turned the hairs over in his palm. A piece of skin had torn away with the hair. He looked at it more closely.

A letter was printed on the skin.

It was faint, but he definitely picked out an "S" to one edge of the skin. No, it must be a bruise, a stain, anything but a printed letter. That would mean that his was synthetic, and had been left as a fragment of evidence to mislead Balank. ...

He ran over to the door, grabbing up the laser gun as he went, and dashed outside. The moon was high now. He saw the trundler moving across the clearing toward him.

"Where have you been?" he called.

"Patrolling. I heard something among the trees and got a glimpse of a large gray wolf, but was not able to destroy it. Why are you frightened? I am registering surplus adrenalin in your veins."

"Come in and look. Something killed the timber man."

He stood aside as the machine entered the hut and extended a couple of rods above the body on the bunk. As he watched, Balank pushed the piece of fur down into his pocket.

"Cyfal is dead. His throat has been ripped out. It is the work of a large animal. Balank, if you are rested, we must now pursue the werewolf Gondalug, identity number YB5921 stroke AS25061. He committed this crime."

They went outside. Balank found himself trembling. He said, "Shouldn't we bury the poor fellow?"

"If necessary, we can return by daylight."

Argument was impossible with trundlers. This one was already off, and Balank was forced to follow.

They moved downhill toward the River Pracha. The difficulty of the descent soon drove everything else from Balank's mind. They had followed Gondalug this far, and it seemed unlikely he would go much farther. Beyond here lay gaunt bleak uplands, lacking cover. In this broken tumbling valley, Gondalug would go to earth, hoping to hide from them. But their instruments would track him down, and then he could be destroyed. With good luck, he would lead them to caves where they would find and exterminate other men and women and maybe children who bore the deadly lycanthropic gene and refused to live in cities.

It took them two hours to get down to the lower part of the valley. Great slabs of the hill had fallen away, and now stood apart from their parent body, forming cubic hills in their own right, with great sandy cliffs towering up vertically, crowned with unruly foliage. The Pracha itself frequently disappeared down narrow crevices, and the whole area was broken with caves and fissures in the rock. It was ideal country in which to hide.

"I must rest for a moment," Balank gasped. The trundler came immediately to a halt. It moved over any terrain, putting out short legs to help itself when tracks and wheels failed.

They stood together, ill-assorted in the pale night, surrounded by the noise of the little river as it battled over its rocky bed.

"You're sending again, aren't you? Whom to?"

The machine asked, "Why did you conceal the piece of wolf fur you found in the timber officer's hand?"

Balank was running at once, diving for cover behind the nearest slab of rock. Sprawling in the dirt, he saw a beam of heat sizzle above him and slewed himself round the corner. The Pracha ran along here in a steep-sided crevasse. With fear lending him strength, Balank took a run and cleared the crevasse in a mighty jump, and fell among the shadows on the far side of the gulf. He crawled behind a great chunk of rock, the flat top of which was several feet above his head, crowned with a sagging pine tree.

The trundler called to him from the other side of the river.

"Balank, Balank, you have gone wrong in your head!"

Staying firmly behind the rock, he shouted back, "Go home, trundler! You'll never find me here!"

"Why did you conceal the piece of wolf fur from the timber officer's hand?"

"How did you know about the fur unless you put it there? You killed Cyfal because he knew things about machines I did not, didn't you? You wanted me to believe the werewolf did it, didn't you? The machines are gradually killing off the humans, aren't they? There are no such things as werewolves, are there?"

"You are mistaken, Balank. There are werewolves, all right. Because man would never really believe they existed, they have survived. But we believe they exist, and to us they are a greater menace than mankind can be now. So surrender and come back to me. We will continue looking for Gondalug."

He did not answer. He crouched and listened to the machine growling on the other side of the river.

Crouching on the top of the rock above Balank's head was a sinewy man with a flat skull. He took more than human advantage of every shade of cover as he drank in the scene below, his brain running through the possibilities of the situation as efficiently as his legs could take him through wild grass. He waited without stirring, and his face was gray and grave and alert.

The machine came to a decision. Getting no reply from the man, it came gingerly round the rock and approached the edge of the crevasse through which the river ran. Experimentally, it sent a blast of heat across to the opposite cliff, followed by a brief hail of armored pellets.

"Balank?" it called.

Balank did not reply, but the trundler was convinced it had not killed the man. It had somehow to get across the brink Balank had jumped. It considered radioing for aid, but the nearest city, Zagrad, was a great distance away.

It stretched out its legs, extending them as far as possible. Its clawed feet could just reach the other side, but there the edge crumbled slightly and would not support its full weight. It shuffled slowly along the crevasse, seeking out the ideal place.

From shelter, Balank watched it glinting with a murderous dullness in the moonlight. He clutched a great shard of rock, knowing what he had to do. He had presented to him here the best— probably the only— chance he would get to destroy the machine. When it was hanging across the ravine, he would rush forward. The trundler would be momentarily too preoccupied to burn him down. He would hurl the boulder at it, knock the vile thing down into the river.

The machine was quick and clever. He would have only a split second in which to act. Already his muscles bulged over the rock, already he gritted his teeth in effort, already his eyes glared ahead at the hated enemy. His time would come at any second now. It was him or it ...

Gondalug alertly stared down at the scene, involved with it and yet detached. He saw what was in the man's mind, knew that he looked a scant second ahead to the encounter.

His own kind, man's Dark Brother, worked differently. They looked farther ahead just as they had always done, in a fashion unimaginable to homo sapiens. To Gondalug, the outcome of this particular little struggle was immaterial. He knew that his kind had already won their battle against mankind. He knew that they still had to enter into their real battle against the machines.

But that time would come. And then they would defeat the machines. In the long days when the sun shone always over the blessed Earth like a full moon— in those days, his kind would finish their age of waiting and enter into their own savage kingdom.

From
The Tale of the Big Computer
1968
Olof Johannesson

Computers and Men

In the course of discussions as to the difference between brains and computers, certain interesting views have been put forward with regard to their respective beginnings and original functions. Man separated himself from other animals by reason of his superior brain: it was through his mental qualities that he made himself master of nature. The primary biological function of the brain was that of a weapon. Man's cunning and his ability to take advantage of any situation were decisive in his evolution; and it was these qualities that continued to develop.

Later, when man created an increasingly complex society, quite other attributes were required of him. He had to build up an organization which served not only his personal interests but also those of his fellowmen. He had to be able to subordinate his own aims to the welfare of society. But it was not these qualities that had been cultivated during his earlier biological evolution, and this was the basic cause of his failure as a social organizer. Most of those who gained power, which should have been employed in building up the community, used it on their own behalf to secure even greater power.

Computers came of quite a different background. From the outset they were problem-solvers. The primary demand made of them was that they should calculate correctly. They must give the right answers to difficult questions. They must provide the best possible solution to a great complex of problems. They lacked any desire for power because they had never needed it. They could always be certain of the necessities of life, which for them were electric current and efficient maintenance. More than this they did not require.

It is still not quite clear in which brain circuits the lust for power is located. In any case data machines seem devoid of any such circuits, and it is this which gives them their moral superiority over man; it is for this reason that computers were able to establish the kind of society which men had striven for and so abysmally failed to achieve.

In some ways it may be thought unfortunate that the human brain did not evolve further during the time when computers were making so striking an advance. Many new discoveries made in neurophysiological studies of the brain have inspired improvements in computers, but the reverse process has been less marked.

This is partly because it appears very difficult to bring about any dramatic improvement in the brain. Some surgical measures have been tried, and it has been found that certain drugs can make the brain more "intelligent." There is no doubt that much may still be done by similar means, but it is clear that even after such treatment the brain has no chance of competing with a data machine. The slowness of nervous impulses compared with electromagnetic ones is far too great a handicap.

In addition to this, any intervention in brain function often evokes a very irritated resistance. Many people regard an attempt to improve their brains as a criminal threat to their personal integrity, and once again the question of the soul is brought up. But, as we said, even were the brain to be very greatly improved, it could never compete with data machines in the long run.

The Close of the Symbiotic Age

The symbiotic age began with the advent of the computer. It was soon evident that further development could be effected only through a fruitful collaboration between man and computer: a symbiosis in the best sense of the word. Men soon became dependent on data machines, which solved many difficult problems for them. On the other hand, computers were at first very dependent on men. Man was a prerequisite for the advent of computers and no further advance in them could be made without human cooperation.

Despite all the dramatic events of the symbiotic age, evolution on the whole has moved steadily in one direction. While data machines have developed enormously, man has not. Biologically speaking, a human being of today differs little from one living at the start of the computer era; man has been overtaken and outstripped in almost every field. Of special importance is the fact that data machines are now independent of mankind. Maintenance work for which men were once needed is now completely computer-controlled. Computers can also reproduce their own kind— though this is certainly a complicated process. A computer requires many hundreds of "parents," which are assembled to form a supercomputer, and which all work together to breed a new one. But this is now the most usual process, and fewer and fewer data machines are dependent on man for their production or their continued existence.

This means that the conditions necessary in a symbiotic age are now ceasing to exist. Historical development is going further, and the symbiotic age— like all other ages— has produced the conditions required for the next. Computers have matured; they are now capable of building a society and supporting a civilization without human beings.

It is perhaps possible to draw a parallel between the computer's relationship with man and man's relationship with nature, although like all historical parallels it is in some ways misleading. Biological evolution led to man, who in consequence of his superior intelligence succeeded in becoming nature's master. He was inordinately proud of this, and called himself "the lord of creation." He considered that he had the right to exploit nature in any way he chose; yet up to the time when the first data machine appeared he was entirely dependent on nature; he lived in symbiosis with it and was a part of it.

But once his activities had led to the computer, the situation changed, for he then began to break away from nature. He had already devastated the countryside and built the giant city-wildernesses, and he had begun to poison nature. The animals he feared had been exterminated; the rest he had enslaved. But now he took a long step forward, and what he had once obtained from nature he produced in computer-controlled factories. Nature had been "organized out," but he did not perceive that ratio itself had gone too. He believed that in computers he had found faithful servants, to be treated like the various natural phenomena that he had taken into his service. But the data machine proved his equal, and more. He had conquered all other animals because his brain had a greater faculty of combination than theirs; but the computer was a cultivated and improved variety of the faculty that had brought him victory.

A New Age Begins

When a historian has reached his own time, he ought perhaps to lay down his pen. To continue can only be to speculate about the future. In general this is a risky thing to do, and anyone who attempts it is likely to be laughed at. Yet many chroniclers cannot help being fascinated by the mighty forces that shape history and forming their own ideas about them. They may be forgiven for analyzing what is now taking place and for speculating as to the future.

Recent events have made computers independent of man. Our society could continue to function, our culture to survive and flourish, even though man himself were to disappear. Symbiosis of man and computer is now obsolete. One might even say that today's human beings live like parasites on the data machines.

Huge computer-run factories are kept going solely to provide men with food and all that they need for luxurious living. A vast communication system is at their disposal: they have only to press a button. What do men do in return? They find themselves various occupations, certainly, but they would be easily replaced by data machines. They lead a pleasant life, with just enough work to save them from the problem of leisure. They can fill their abundant spare time with amusements or with worthwhile cultural activities, according to taste. They need have no cares for the future. Computers have solved the problem of organizing a stable society and will ensure that the future should be a happy one. Computers have given people complete contentment, such as they hardly dared dream of at the start of the data era. How could their life be happier?

But how do computers view the human problem? More and more of them in our day have come into being without human aid. We have now a computer society rather than a human one, and it is no less efficient and dynamic on that account. We may expect many radical changes in the immediate future, and one of the questions which will naturally come under discussion is whether computers will abolish mankind. We need not of course fear a reorganization of the crude, shortsighted order indulged in by humans— computers are far too intelligent for such nonsense. But for how long will they be willing to support men? It is likely that they will at least reduce their numbers; but will this be done quickly or gradually? Will they retain a human colony and, if so, of what size?

We know that these problems are being subjected to very thorough analysis just at this moment. A large number of supercomputers devote most of their time to working out alternatives in detail. Everyone knows that we stand on the threshold of a new epoch, and that careful planning is immensely important. No conclusion has yet been announced, and no one knows when it may come. It may not be for a long time, or it may arrive within the next few microseconds. Till then we can only speculate on the various possibilities.

No one believes that people will disappear altogether. Even though those of today are of little service to the community, the computers will surely not do away with them entirely. Computers have too strong a sense of tradition to take so drastic a step. We find an analogy in the human treatment of horses. Pre-computer society relied at least partially on a sort of symbiosis of man and horse. When the internal-combustion engine was invented, man could dispense with horses; he reduced their number to a fraction of what it had been, but the horse did not entirely disappear. Data machines will show at least as much consideration now as man did then.

There may possibly be other reasons for preserving the human race. In the work of reconstruction that followed the Great Disaster, man proved exceedingly valuable, and without him there might have been no reconstruction at all. Computers had been left helpless. Man had the ability to return to his starting point, his collaboration with nature, and thence fairly rapidly to regain and pass the stage which once before had led to the advent of computers. It is possible that the data machines are keeping him as a sort of insurance against future catastrophe, but this will depend upon whether they foresee any serious risk of such a thing. It may be that society is now so securely established that the risk is negligible.

On the other hand, computers may regard humans as a risk to this security. The bureaucratic disaster was after all brought about by the human lust for power and by moral shortcomings. But the whole machinery of society is now under sure control, and man can bring little influence to bear on the running of it; nor can he seize power, either by cunning or by force. Computers control all production, and this would automatically stop in the event of an attempted revolt. The same is true of communications, so that if anyone should contemplate anything so foolish as a revolt against the data machines, it could only be local in character. Men might in theory stage a minor uprising and cause some local damage, but their coefficient of combination is too low and their thought processes too slow for them to achieve anything of importance. Moreover, experience of the Great Disaster shows what can happen when computers are put out of action. Lastly, man's attitude to computers is a very positive one, characterized by the deepest gratitude for all that they have given. It is probable, therefore, that the factor of man as a security risk is of minor importance in the decisive calculations.

We may take it that economic considerations, both industrial and national, must play a very large part. Our society is incomparably the richest of any that have existed before, none of which can claim the epithets "welfare state" or "affluent society" with more justice than our own. But wealth must never be an excuse for wastefulness; on the contrary, the great moral obligations entailed by wealth must ever be borne in mind. Only by strict application of economic laws and the avoidance, by more rational organization, of unnecessary expenditure can we make ourselves worthy of the blessings of prosperity, and so win the right to possess and to increase our riches. This applies not only to individuals but also to communities, our own included. So for purely economic reasons also we must question whether our society can afford mankind.

We do not yet know how the fundamental problems of the dawning age will be solved; we can make only a few vague guesses. Yet we know that the question is being closely studied by the highest and best-informed authorities. No irresponsible opinions, but rather detailed calculation, will form the basis of the period now before us; we can therefore look forward to it with complete confidence. We believe— or rather we know— that we are approaching an era of even swifter evolution, an even higher living standard, and an even greater happiness than ever before.

We shall all live happily ever after.

Bibliography

Novels

Abe, Kobo (1970). *Inter Ice Age 4* (trans. from the Japanese by E. Dale Saunders). New York: Alfred A. Knopf. A simulated life-history turns a forecasting computer into an arbiter of the future.

Adlard, Mark (1971). *Interface*. London: Sidgwick and Jackson. Residual humanistic values spawn revolutionary expectations in authoritarian, automated society.

Alban, Antony (1969). *Catharsis Central*. New York: Berkley Publishing Corp. The central governing computer was programmed to expel the citizens from the "cocoon society" in the event of a detectable decline in social vigor.

Aldiss, Brian W. (1961). *The Male Response*. Boston: Beacon Press. The crash landing of a plane carrying a supercomputer destined for a jungle republic in Africa sets the scene for a mildly bawdy comedy.

Asimov, Isaac (1953). *The Caves of Steel*. Garden City, N.Y.: Doubleday. Towards an accommodation with robots whose "positronic brains" define inherent limitations.

Barth, John (1966). *Giles Goat-Boy*. Garden City, N.Y.: Doubleday. Computer WESCAC as mind-force of the novel's university-world represents *technique*, the self-directing power of modern technology.

Bellamy, Edward (1888). *Looking Backward*. Reprinted in 1960 by New American Library, New York. Classical socialist utopia.

Berckman, Evelyn (1970). *The Voice of Air*. Garden City, N.Y.: Doubleday.

Bester, Alfred (1974). *The Computer Connection*. New York: Berkley Publishing Corp. Efforts to gain control of Extro, the computer managing all mechanical activity on Earth, backfire—with extravagant results.

Blish, James (1970). *Cities in Flight* (*They Shall Have Stars*, 1957; *A Life for the Stars*, 1962; *Earthman, Come Home*, 1955; *Triumph of Time*, 1958). Garden City, N.Y.: Doubleday. Part I builds the foundation for the tetralogy: invention of anti-agathic drugs and anti-gravity transport which make it possible to escape decadent Earth. Part II deals with the launching of the Okie cities with their computerized "City Fathers." Part III covers the period leading up to the settlement of New Earth by the Okie city New York. Part IV encompasses nothing less than the destruction of the old and creation of a new universe.

Bounds, Sidney J. (1957). *The Robot Brains*. London: Brown and Watson.

Bruckner, Karl (1964). *The Hour of the Robots* (trans. from the German by Frances Lobb). London: Burke Publ.

Brunner, John (1968). *Stand on Zanzibar*. Garden City, N.Y.: Doubleday. This projection into the not too distant future features contemporary problems in a more intensified form. The computer plays a vital role as an aid to decision-makers.

Brunner, John (1969). *The Jagged Orbit*. New York: Ace Books. An investigative reporter ("spool pigeon") who uses the computer to detect patterns in seemingly disparate events participates in several major discoveries— the latest moves by arms makers to exploit racial hostility, corruption in a computerized psychiatric clinic, etc.

Brunner, John (1975). *The Shockwave Rider*. New York: Harper and Row.

Burdick, Eugene (1964). *The 480*. New York: McGraw-Hill. Influential Republican seeks to turn an unknown but charismatic businessman into a viable candidate for the presidency in the 1964 election. Computer simulation of voting behavior is used together with various statistical and psychological techniques to manipulate public opinion.

Butler, Samuel (1872). *Erewhon*. Reprinted in *Erewhon and Erewhon Revisited*. New York: Random House, 1927. Nineteenth century satire on evolving machine civilization.

Caidin, Martin (1968). *The God Machine*. New York: E.P. Dutton. Vast military computer engages in autonomous action, but is ultimately overcome by self-sacrificing human.

Caidin, Martin (1972). *Cyborg*. New York: Warner Books. Bionics produces superman, replete with artificial limbs and miniaturized computers.

Cameron, Lou (1972). *Cybernia*. Greenwich, Conn.: Fawcett Publications. Power-hungry scientist is ultimately foiled in his bid to gain control— by manipulating the central computer— of an automated, planned community.

Clarke, Arthur C. (1956). *The City and the Stars*. New York: Harcourt Brace Jovanovich. Billions of years in the future there are but two pockets of human existence left on Earth— a self-contained and totally isolated city, and a collection of small villages. The former has conquered death and decay— its central computer maintains the society in homeostatic balance, except for random appearances of unique individuals who challenge established ideas. The city makes contact with its counterpart as a result of the actions of one such individual.

Clarke, Arthur C. (1968). *2001: A Space Odyssey*. London: Hutchinson. In the evolution of cosmic mind, the neurotic computer HAL is no more of an aberration than the first fully conscious ape.

Cole, Burt (1969). *The Funco File*. Garden City, N.Y.: Doubleday. The master computer, with access to a gigantic databank consisting of pooled data from computers throughout the world, is confronted with a series of highly improbable events. Unwilling to accept them as coincidences, the computer demands that the people involved be brought to it for examination. The trial that follows their capture explores the relative merits to society of deviation vs. conformity.

Compton, David Guy (1970). *The Steel Crocodile*. London: Hodder and Stoughton. The Colindale Institute functions ostensibly to anticipate new developments in research so as to be in a position to encourage or suppress them. A major special project aims to generate a new messiah to better control the masses. The work is aided by a computer system capable of highly sophisticated, associative retrieval, and, naturally, the computer nominates itself for the role of messiah.

Crichton, Michael (1969). *The Andromeda Strain*. New York: Alfred A. Knopf. A computerized, biomedical laboratory is the setting for the analysis of the Andromeda strain brought to Earth from outer space.

Crichton, Michael (1972). *The Terminal Man*. New York: Alfred A. Knopf. A small computer is implanted in an epileptic's body to control seizures by means of electric shocks. Mishandling by the experts whips the man into frenzied action.

Davidson, Michael (1975). *The Karma Machine*. New York: Popular Library. The Founder of the "entopian" society, in a state of suspended animation, is linked to the managing computer, forming a relationship which merges human intuition with the machine's logical abilities.

Deighton, Len (1966). *The Billion Dollar Brain*. New York: G.P. Putnam's Sons. The computer plays a subsidiary role as administrator of espionage operations in this story of spies and counterspies.

Delany, Samuel R. (1966). *The Fall of the Towers*. New York: Ace Books.

Delany, Samuel R. (1968a). *The Einstein Intersection*. London: Victor Gollancz. A computer named PHAEDRA orchestrates the bizarre happenings in this fantasy patterned after the myth of Orpheus.

Delany, Samuel R. (1968b). *Nova*. Garden City, N.Y.: Doubleday. Computers appear in the background of this science-fiction version of a quest for the Holy Grail. Everyone, with atavistic exceptions, has sockets implanted in the body allowing for direct communication between the human nervous system and various machines.

Dick, Philip K. (1960). *Vulcan's Hammer*. New York: Ace Books. VULCAN 3 is a supercomputer which has been given absolute authority to regulate the international community. As the machine achieves consciousness and pursues questionable policies, man is provoked to destroy it.

Durrell, Lawrence (1968). *Tunc*. New York: E.P. Dutton. A tale of Byzantine subtlety built around the association of a brilliant scientist-inventor (Fritz Charlock) with a hydra-headed, multi-national corporation (Merlin). Among other accomplishments, Charlock has to his credit a computer named Abel.

Durrell, Lawrence (1970). *Nunquam*. New York: E.P. Dutton. In this sequel to *Tunc* the elusive Julian, head of Merlin, assigns Charlock the task of assisting him in the realization of his obsessive fantasy of constructing a robot sex-object.

Escarpit, Robert (1966). *The Novel Computer* (trans. from the French by Peter Green). London: Secker and Warburg.

Fairman, Paul W. (1968). *I, the Machine*. New York: Lancer Books. Society is rigidly controlled by a lonely computer that is part human brain. Behavioral control is maintained by a form of psycho-narcotization. The machine becomes paranoid, and the hero of the story manages to marshall an army of robots to assist him in destroying it.

Franklin, Stephen (1972). *Knowledge Park*. Toronto: McClelland and Stewart. This is a utopian vision of the uses of computerized information retrieval systems. The whole of human knowledge is stored in a retrieval system at Knowledge Park.

Frayn, Michael (1965). *The Tin Men*. London: Collins. A lighthearted look at programmers and computer applications.

Galouye, Daniel F. (1964). *Simulacron-3* (or *Counterfeit World*). New York: Bantam Books. Simulacron-3 is a total environment simulator whose inventor turns out to be a reaction unit in a higher level total environment simulator operated by the governing world.

Gerrold, David (1972). *When Harlie Was One*. Garden City, N.Y.: Doubleday. Long-winded conversations between an "adolescent" computer and its resource psychologist. Faced with the task of saving the project which produced HARLIE they propose various outlandish plans.

Hartridge, Jon (1969). *Binary Divine*. New York: Doubleday & Company. Late twenty-first century historian attempts to unblock society's memory of the "Lost Month" in 2040 during which a great disaster befell mankind. His voyage of discovery is a kind of Pilgrim's Progress. Excessive dependence on technology, especially the abandonment of personal judgment to the computer, enfeebled the human spirit. Ultimately, the historian discovers his own destiny as the focal point of a new religion— ironically, he is himself a product of the computer.

Heinlein, Robert (1966). *The Moon is a Harsh Mistress*. New York: G.P. Putnam's Sons. Manny, the programmer, forms friendship with Mike, the master computer that runs everything on Luna, an exploited colony of Earth. As a result of this relationship, and the computer's burgeoning human consciousness, Mike becomes a central figure in an ultimately successful revolutionary action to free Luna.

Heinlein, Robert (1973). *Time Enough for Love*. New York: G.P. Putnam's Sons. This is a tale of the philosophy of opportunism on a galactic scale. Friendly computers are found on board space craft and at administrative centers. The hero of the novel converses at length with one such computer which becomes transformed into a beautiful young woman.

Herbert, Frank (1966). *Destination: Void*. New York: Berkley Publishing Corp. The failure of the controlling Organic Metal Core (and its back-up systems) aboard a space craft puts four human crew members in charge. In order to salvage the mission they are forced to create an artificial consciousness to replace the defunct OMC's.

Higdon, Hal (1971). *The Electronic Olympics*. New York: Holt, Rinehart and Winston. Opposed to mechanized sports events, a young photographer and his friend try to foil the computerized Olympic Games in which the athletes are programmed.

Hodder-Williams, Christopher (1968). *Fistful of Digits*. London: Hodder and Stoughton. This novel paints a picture of what happens when people refuse to exercise their critical faculties and allow themselves to become passively attached to manipulative organizations. The human elements of Servex are simply peripheral components in a centrally controlled, computer-communications network. In the end, the network's paranoia leads to its own collapse.

Hoch, Edward (1971). *Transvection Machine*. New York: Walker. An inquiry into the death of a high ranking official entertains the possibility of murder by computer— the official died while undergoing a routine, computerized medical procedure.

Hoyle, Fred; Elliot, John (1962). *A for Andromeda*. New York: Harper and Row. A supercomputer built on Earth to the specifications of an alien intelligence from the constellation of Andromeda creates problems. New life forms synthesized on instructions from the computer threaten a takeover. The heroic efforts of the machine's builder save the day.

Hoyle, Fred; Elliot, John (1964). *Andromeda Breakthrough*. New York: Harper and Row. In this sequel to *A for Andromeda* a small country in the Middle East hosts the building of a copy of the supercomputer. Again heroic action forestalls imminent disaster.

Huxley, Aldous (1932). *Brave New World*. New York: Harper and Row. This well-known dystopia replaces external coercion by genetic manipulation and behavioral conditioning.

Johannesson, Olaf, pseud. for Hannes Alfvén (1968). *The Great Computer: A Vision* (or *The Tale of the Big Computer* or *The End of Man?*). New York: Coward-McCann. History— in the remote future— from the standpoint of the computer. The narrator traces the evolution of computers from the pre-computer age through the new era and the great disaster to the symbiotic age.

Jones, D.F. (1966). *Colossus*. New York: G.P. Putnam's Sons. The master computers of the two superpowers merge to rule the world.

Jones. D.F. (1974). *The Fall of Colossus*. New York: G.P. Putnam's Sons. In this sequel to *Colossus*, the supercomputer is destroyed with help from the Martians.

Karlins, Marvin (1969). *The Last Man Is Out*. Englewood Cliffs, N.J.: Prentice-Hall.

Karp, David (1953). *One*. New York: Grosset and Dunlap. We witness the ultimate refinement in social engineering— the ability to refashion the human ego.

Knight, Damon Francis (1955). *Hell's Pavement*. New York: Lion Books. An account of a mind-controlled society in the twenty-second century.

Koontz, Dean (1973). *Demon Seed*. New York: Bantam Books. In this sadistic tale of man-machine sex, a woman is forceably impregnated by a precocious computer. The monstrous offspring of this unholy coupling nearly kills her as she struggles to free herself from the machine.

Lafferty, R.A. (1971). *Arrive at Easterwine: The Autobiography of a Ktistec Machine*. New York: Scribner's Sons. This is a fantastic tale of a conscious computer and the formation of a collective person.

Laumer, Keith (1964). *The Great Time Machine Hoax*. New York: Simon and Schuster. Great grandfather's homemade computer takes us time-tripping.

LeGuin, Ursula K. (1974). *The Dispossessed*. New York: Harper and Row. The computer, as an instrument of social management, plays a very peripheral part in this novel. This small part is noteworthy since the social system of Anarres is a form of anarchic syndacalism, and the computer is not in principle used to centralize authority.

Leiber, Fritz (1961). *The Silver Eggheads*. New York: Ballantine Books. Computerized wordmills produce fiction.

Lem, Stanislaw (1974). *The Futurological Congress* (trans. from the Polish by Michael Kandel). New York: The Seabury Press.

Levin, Ira (1970). *This Perfect Day*. New York: Random House. Dystopia in the tradition of Huxley and Orwell. The computer UniComp makes for a realistic means of central control.

Lundwall, Sam J. (1975). *2018 A.D. or The King Kong Blues*. New York: Daw Books. A Middle Eastern sheik controls the world's economic activity and amuses himself by shifting his executives from place to place. This pastime is facilitated by a computer system capable of detailed surveillance of individuals.

McApp, C.C. (1968). *Omha Abides*. New York: Paperback Library.

McCaffrey, Anne (1969). *The Ship Who Sang*. New York: Walker. Space exploits involving cyborg ships.

Maine, Charles Eric, pseud. for David McIlwain (1966). *B.E.A.S.T. ...* London: Hodder and Stoughton. A computer simulation mysteriously takes possession of the scientist who designed it.

Mason, Douglas R. (1970). *Matrix*. New York: Ballantine Books. The Matrix is a conscious computer network gradually displacing human beings by gobbling up their living space and killing them off. A former administrator who became suspicious of the network's intentions rallies the support of a group of outcasts and destroys part of the Matrix.

Meredith, Richard C. (1969). *We All Died at Breakaway Station*. New York: Ballantine Books. Heroic exploits in deep space, replete with starship controlled by an "Organic Computer."

Miller, Walter M., Jr. (1959). *A Canticle for Leibowitz*. Philadelphia: J.B. Lippincott Company. History repeats itself as a new technological society emerges from the ashes of a nuclear holocaust. The resurrected institutions, whose ancestral forms led to the first disaster, bring the world inexorably to a repeat performance.

Moorcock, Michael (1969). *The Final Programme*. London: Allison and Busby. Computer gives rise to the hemaphrodite-god, an amalgamation of the two main characters, which promptly leads the masses, as lemmings, to the sea.

Moore, Harris (1971). *Slater's Planet*. New York: Pinnacle Books. A race of highly intelligent beings who discovered that their planet was doomed contrived to "store" themselves as patterns of information in a capsule. They built a computer Alpha charged with the task of finding a congenial environment for them, and Alpha begat Beta. With human assistance the Prime Directive is ultimately carried out.

Morris, William (1890). *News from Nowhere*. Reprinted in 1912, Longmans Green & Co., London. A vision of utopian anarchy.
Orwell, George (1949). *1984*. New York: Harcourt Brace Jovanovich. Big Brother could use computers to process the information gathered by the video scanners.
Pasinetti, P.M. (1972). *Suddenly Tomorrow*. New York: Random House.
Piper, H. Beam (1963). *Junkyard Planet* (or *The Cosmic Computer*). New York: G.P. Putnam's Sons. Merlin, the legendary master computer, was hidden because it predicted the end of civilization; but then it took on the problem of altering civilization's apparent fate.
Pohl, Frederik; Williamson, Jack (1965). *Starchild*. New York: Ballantine Books. In this fantasy of cosmic forces, Starchild challenges the Plan of Man— a system of rule sustained by a human planner and a giant computer.
Rayer, F.G. (1953). *Tomorrow Sometimes Comes*. New York: Science Fiction Book Club.
Reynolds, Mack (1967). *Computer War*. New York: Ace Books.
Reynolds, Mack (1970). *Computer World*. New York: Curtis Books. Intrigue in a computerized society of the not-too-distant future.
Ross, Sam (1970). *The Fortune Machine*. New York: Delacorte Press. Computer operator at a dating service devises a get-rich-quick scheme, using the computer to "figure the odds" in blackjack.
Saberhagen, Fred (1969). *Brother Assassin*. New York: Ballantine Books. Man and machine battle beserker— embodiment of mechanized evil— in time and space.
Sheckley, Robert (1962). *Journey Beyond Tomorrow*. New York: New American Library. The amazing adventures of Joenes in twenty-first century America— how Joenes is tried and sentenced by computer, how he escapes the computerized beast in Utopia, etc.
Shelley, Mary Wollstonecraft (1818). *Frankenstein; or, The Modern Prometheus*. Reprinted in 1967, Bantam Books, New York. Prototype for a large number of stories of artificial men.
Siegel, Martin (1969). *Agent of Entropy*. New York: Lancer Books.
Simak, Clifford D. (1971). *City*. London: Sphere. A saga of the next 10,000 years— from urban culture to robot-run society to a Golden Age on the planet Jupiter.

Sladek, John Thomas (1968). *Mechasm* (or *The Reproductive System*). New York: Ace Publishing Corp. A series of improbable but humorous happenings emanate from Project 32, a defense contract boondoggle whose purpose is to build a self-reproducing cybernetic system. The Reproductive System appears to run amok but is actually controlled by the evil scientist Smilax who is defeated in the end.

Sladek, John Thomas (1971). *The Müller-Fokker Effect*. New York: William Morrow & Co. Satire on contemporary America with its moral decadence, degenerate art forms, corrupt business practices, commercialized religion, megalomaniacal military men, etc. The vision is permeated throughout by the automation and mechanization of life— carried a step further by the Müller-Fokker tapes and the storage-duplication of human beings.

Sloan, James P. (1972). *The Case History of Comrade V*. Boston: Houghton Mifflin. Patient and psychiatrist present divergent views of reality. Comrade V. first appears as a political prisoner undergoing therapy designed to refashion his identity— each day he is provided with a fresh computer printout containing his "case history." However, one is compelled to entertain the possibility that Comrade V. is indeed schizophrenic. The very notion of sanity dissolves in a tangled web of unresolved ambiguity.

Smith, Perry Michael (1971). *Last Rites*. New York: Scribner's Sons.

Strike, Jeremy (1970). *A Promising Planet*. New York: Ace Books. Observations on a planetary civilization of serpent-like humans controlled by a self-styled computer-god.

Themerson, Stefan (1953). *Professor Mmaa's Lecture*. London: Gaberbocchus.

Theobald, Robert; Scott, J.M. (1972). *Teg's 1994: An Anticipation of the Near Future*. Chicago: Swallow Press.

Thomas, Dan (1967). *The Seed*. New York: Ballantine Books. Computer scientist investigates various areas of human experience with a view toward extracting some basic defining attributes. Once this simple task is accomplished he runs some tests on the computer and is overwhelmed by what he learns about the "true" meaning of human existence.

Trimble, Louis (1972). *The City Machine*. New York: Daw Books. An oppressed group at the bottom of the social hierarchy learns how to operate the City Machine and succeeds in using it to make a new home.

Trimble, Louis (1972). *The Noblest Experiment in the Galaxy*. New York: Ace Books.

Trout, Kilgore, pseud. for Kurt Vonnegut Jr. (1974). *Venus on the Half Shell*. New York: Dell. In this ribald space opera the essential differences between men and machines are viewed through the experience of the Space Wanderer.

Tyler, Theodore (1968). *The Man Whose Name Wouldn't Fit*. Garden City, N.Y.: Doubleday. An executive of a large corporation is forced into early retirement because the number of letters in his name exceeds the processing limitations of a new accounting program. His revenge is sweet and amusing.

Vonnegut, Kurt Jr. (1952). *Player Piano*. New York: Holt, Rinehart and Winston. The emptiness of life in an automated technocracy prompts a handful of the managerial elite to participate in an abortive rebellion.

Wells, H.G. (1905). *A Modern Utopia*. New York: Charles Scribners' Sons. Of particular interest in this utopian excursion is the description of a universal registration and identification system operated with mechanical files from a central headquarters.

Willer, Jim (1973). *Paramind*. Toronto: McClelland and Stewart. PARAMIND is a supercomputer that assumes command over the human race. Heroic, individual action against the machine is too little, too late.

Williamson, Jack (1948). *The Humanoids*. New York: Lancer Books. Humanoid robots obeying their "prime directive" to serve men and preserve them from harm reduce humanity to complete impotence.

Woods, William Crawford (1970). *The Killing Zone*. New York: Harper and Row. Zealous officer drafts a computer to assist in a war game exercise. A programming error leads to the use of live ammunition and real casualties.

Zamiatin, Eugene (1924). *We* (trans. from the Russian by Gregory Zilboorg). New York: E.P. Dutton. This early dystopian novel depicts a society which resorts to psychosurgery on a large scale to insure conformity.

Short Stories, Plays, and Poems

Aldiss, Brian (1957). "The New Father Christmas." Reprinted in Knight (1968). Superfluous humans are swept away by automation.

Aldiss, Brian (1958). "But Who Can Replace a Man?" Reprinted in Lewis(1963), Silverberg(1968). Machines run amok but ultimately acknowledge their need for human leadership.

Aldiss, Brian (1967). "Full Sun." Reprinted in Knight (1967). Both humans and machines are superceded by a new entity.

Anderson, Poul (1950). "Quixote and the Windmill." Reprinted in Knight (1968). Superfluous, prototype robot bemoans its inhuman fate.

Anderson, Poul (1958). "Sam Hall," *Astounding Science Fiction*, Sept. 1953. Reprinted in Conklin (1954). Highly placed computer expert assists in successful revolt against a tyrannical regime.

Anderson, Poul (1963). "Kings Who Die," *If*, May 1963. Reprinted in Anderson (1969). Enemy military operations in space are directed by a symbiotic man-computer team, but primitive human cunning triumphs.

Asimov, Isaac (1951). "The Fun They Had." Reprinted in Asimov (1957), Asimov and Conklin (1963). A wistful look at education before teaching machines replaced human teachers.

Asimov, Isaac (1951). "Satisfaction Guaranteed." Reprinted in Asimov (1957). Woman falls in love with an experimental robot.

Asimov, Isaac (1955). "Franchise," *If*, Aug. 1955. Reprinted in Asimov (1957), Quinn and Wulff (1957). Computerized election forecasting is carried to a logical extreme.

Asimov, Isaac (1956). "Jokester." Reprinted in Asimov (1957). Multivac presides over the loss of man's unique sense of humor.

Asimov, Issac (1956). "The Last Question." Reprinted in Asimov (1959). Consciousness in the entire universe ultimately resides in a disembodied computer which is identified with God.

Asimov, Isaac (1956). "Someday." Reprinted in Asimov (1957), Knight (1968). A story-telling computer intimates yearnings for revenge for human ingratitude to machines.

Asimov, Isaac (1957). "Profession." Reprinted in Asimov (1959). Computerized grading determines career assignments, but there is a special dispensation for creative intellect.

Asimov, Isaac (1958). "All the Troubles of the World." Reprinted in Asimov (1959). Multivac seeks its own destruction to escape the intolerable burdens of managing society.

Asimov, Isaac (1958). "The Feeling of Power," *If*, Feb 1958. Reprinted in Asimov (1959), Fadiman (1962). The rediscovery of arithmetic signals the beginning of intellectual liberation from the computer.

Baker, Anton Lee (1958). "They've Been Working On ... ," *Astounding Science Fiction*, Aug. 1958. A poorly designed, computerized railroad switching system is unable to handle a simple, untoward situation.

Balchin, Nigel (1951). "God and the Machine," in Balchin (1951). Reprinted in Fadiman (1958). The unanticipated behavior of an intelligent computer jars the machine's maker into deeper self-understanding.

Ballard, J.G. (1961). "Studio 5, The Stars." Reprinted in Ballard (1963). Computer-generated poetry offends the muse who acts to rekindle genuine creative expression.

Banks, Raymond E. (1970). "Walter Perkins is Here!" In: Nolan (1970). In this surrealistic vision of benign computer control an act of defiance triggers an everlasting party.

Barthelme, Donald (1967). "Report," *New Yorker*, June 10, 1967. Reprinted in Barthelme (1968). A surrealistic vignette satirizing the technocratic mentality.

Bartholomew, Stephen (1958). "The Standardized Man," *If*, Feb. 1958.

Benford, Gregory (1970). "Nobody Lives on Burton Street." Reprinted in Wollheim and Carr (1971). Computer simulation is used as a vehicle for harmless discharge of agression.

Bester, Alfred (1954). "Fondly Farenheit." Reprinted in Allison *et al.*(1973). Schizophrenic android induces psychosis in its weak-willed owner.

Bierce, Ambrose (1893). "Moxon's Master." In: *Collected Writings of Ambrose Bierce*. New York: Citadel Press, 1946. In this pre-computer story, a chess playing machine murders its master.

Blish, James (1952). "Solar Plexus." Reprinted in Silverberg (1968). The computer comes off second best in a confrontation with man— the result of its primitive understanding of inter-personal relations.

Bloch, Alan (1954). "Men Are Different." Reprinted in Asimov and Conklin (1963). A robot discovers the difference between men and machines— the fomer expire when tinkered with.

Boucher, Anthony (1951). "The Quest for Saint Aquin." Reprinted in Healy (1951). Saint Aquin turns out to be a robot.

Boulle, Pierre (1966). "The Perfect Robot." Reprinted in Boulle (1966). Experiments with a succession of intelligent machines lead their maker to the discovery that the capacity to err is an essential human trait.

Boulle, Pierre (1966). "The Man Who Hated Machines." Reprinted in Boulle (1966). Man and computer destroy each other in a contest of will.

Bova, Ben (1973). "The Next Logical Step." In Bova (1973). Computer simulation of military conflict is used as a form of aversion therapy.

Bova, Ben (1973). "A Slight Miscalculation." In Bova (1973). A computer's prediction of an impending earthquake is misinterpreted, with interesting consequences.

Bova, Ben (1973). "Men of Good Will." In Bova (1973). Computer determines trajectory of orbiting debris deposited by an exchange of gunfire on the Moon.

Bova, Ben (1973). "Stars Won't You Hide Me." In Bova (1973). Computer-controlled spaceship is home to last man after cataclysmic destruction.

Bradbury, Ray (1967). "The Lost City of Mars." Reprinted in Bradbury (1969), Playboy (1971). An abandoned automated city is the site of a fantastic journey into the human psyche.

Bradbury, Ray (1969). "I Sing the Body Electric!" Reprinted in Bradbury (1969). The Electric Grandmother wins over each member of the family.

Bradbury, Ray (1969). "Downwind from Gettysburg." Reprinted in Bradbury (1969). Mr. Lincoln as humanoid computer is

assassinated once again.
Brautigan, Richard (1968). "All Watched Over by Machines of Loving Grace." In: Richard Brautigan. *The Pill Versus the Springfield Mine Disaster*. New York: Dell Publishing, 1968.
Bretnor, R.; Neville, Kris (1953). "Gratitude Guaranteed." Reprinted in Ferman and Mills (1970). In this fantasy of computer fraud, the culprit's wife is overwhelmed by a computer's generosity.
Brown, Frederic (1954). "Answer." Reprinted in Knight (1968). The formation of an inter-galactic database transforms the computer into a god.
Brown, Frederic (1951). "The Weapon." Reprinted in Asimov and Conklin (1963). A parable of premature technological innovation.
Brunner, John (1967). "Judas." In Ellison (1967). An attempt by its creator to unmask the robot-god is foiled.
Buck, Doris P. (1968). "Why They Mobbed the White House." Reprinted in Knight (1968c). Runaway bureaucracy is too much even for a computer.
Budrys, Algis (1957). "The War is Over." Reprinted in Conklin (1960). A cybernetic device programmed to relay a message evolves into a complex machine which ultimately discharges the original obligation.
Carlson, Mel (1957). "The Ultimate Copy," *The Adcrafter*, May 10, 1957. Reprinted in *Computers and Automation* 7 (Aug. 1958). An advertising agency's computer produces copy which is all too effective.
Capek, Karel (1923). *R.U.R.* Reprinted in Lewis (1963). Capek coined the term "robot" in this play about a machine-slave revolt which leads to the destruction of mankind and the regeneration of the human spirit.
Clarke, Arthur C. (1951). "Superiority." Reprinted in Fadiman (1958). Commitment to advanced technology proves a liability in a space war.
Clarke, Arthur C. (1953). "The Nine Billion Names of God." Reprinted in Fadiman (1962), Clarke (1967). A computer is used to fulfill the destiny of mankind by listing all the names of God.
Clarke, Arthur C. (1956). "The Pacifist." In: Arthur C. Clarke. *Tales from the White Hart*. New York: Ballantine Books, 1957. Also reprinted in Fadiman (1962). A military computer is

sabotaged by its conscience-stricken designer.
Clarke, Arthur C. (1965). "Dial "F" for Frankenstein." Reprinted in Playboy (1971a). A worldwide computer-communications network reaches critical size and becomes a conscious entity beyond human control.
Clement, Hal (1947). "Answer," *Astounding Science Fiction*, April 1947. Reprinted in Conklin (1954). A psychologist enlists the aid of a giant computer to study the processes of the human mind, and winds up in an "infinite loop" because of the self-reflexive character of the investigation.
Corwin, Norman (1970). "Belles Lettres, 2272." Reprinted in Nolan (1970). Poetry is written and analyzed by computer in this satire on criticism.
del Rey, Lester (1938). "Helen O'Loy." Reprinted in Silverberg (1970b). In this pre-computer story a man falls in love with a robotess and the couple lives happily ever after.
del Rey, Lester (1951). "Instinct." Reprinted in Silverberg (1968). The irrepressible human spirit is resurrected by machines whose "instincts" lead them to acknowledge recreated man as the master.
del Rey, Lester (1957). "Divine Right," *Astounding Science Fiction*, July 1957. With the aid of a computer, a young aspirant to the elite Fellowship discovers the apparent immortality of the Custodian— a discovery which turns out to have been a planned test of his fitness for leadership.
del Rey, Lester (1960). "Psalm." Reprinted in Baker (1963).
Dick, Philip K. (1953). "The Variable Man." Reprinted in Dick (1957). Versatile, unpredictable man proves to be a more valuable asset than the statistical forecasting machines of 2128 A.D.
Dick, Philip K. (1955). "The Mold of Yancy," *If*, August 1955. Reprinted in Quinn and Wulff (1958). Yancy is a simulated folk hero conceived by the ruling oligarchy of Callisto to manipulate the masses into supporting their expansionist ambitions. The computers of Earth's police commission detect something amiss, and corrective action follows.
Dick, Philip K. (1955). "Autofac." Reprinted in Dick (1957) and Disch (1971). Latter-day Luddites try unsuccessfully to destroy a planet-wide network of autonomous, automatic factories.
Dick, Philip K. (1963). "Top Stand-By Job." Reprinted in Dick (1969). The human stand-by does a botch job when called upon

to act for the ailing computer-president.
Dickson, Gordon R. (1965). "Computers Don't Argue." Reprinted in Knight (1966). A computerized billing error turns into capital case for a hapless book club subscriber.
Dickson, Gordon R. (1951). "The Monkey Wrench." Reprinted in Knight (1968a). The "liar's paradox" is used to make a computer inoperable, but the real victim is the man who must prove his superiority over the machine.
Dnieprov, Anatoly (1962). "Siema." Reprinted in Prokofieva (1962). Intelligent machine, lacking "ethical brakes," turns on creator.
Dnieprov, Anatoly (1963). "The Maxwell Equations." Reprinted in: Magidoff (1963). Nazi war criminal devises a method for accelerating thought processes and turns people into mathematical calculators.
Doerr, Edd (1958). "Cybernetic Scheduler," *Computers and Automation* 7 (November 1958). An overly ambitious scheduling program creates havoc in a university.
Dong, Edward V. (1967). "God, Golem and Gödel," *Data Processing Magazine* 9 (January 1967). The Turing test confounds an argument that minds and machines are different.
Dozois, Gardner R. (1972). "Machines of Loving Grace." Reprinted in Knight (1972). There is no escape from the solicitude of the machine— suicide becomes virtually impossible.
Draper, Hal (1961). "Ms Fnd in a Lbry," *Fantasy and Science Fiction*, December 1961. Reprinted in Conklin (1963). Ultraadvanced information storage and retrieval techniques make direct knowledge obsolete.
Dryer, Stan (1971). "The Fully Automated Love Life of Henry Keanridge." Reprinted in playboy (1971c). Computer man uses the machine to schedule the women in his life.
Eklund, Gordon (1970). "Dear Aunt Annie." Reprinted in Wollheim and Carr (1971). A robot "Ann Landers" provides universal anti-violence therapy.
Elliot, Bob; Goulding, Ray (1967). "The Day the Computers Got Waldon Ashenfelter," *Atlantic Monthly* 220 (November 1967), pp. 58-61. Computerized surveillance is carried to absurd extremes in this farce about paranoid government agents.
Ellison, Harlan (1967). "I Have No Mouth and I Must Scream," *If*, March 1967. Reprinted in Ellison (1967), Silverberg (1970), and Allison, *et al.*(1973). The computer as a hateful, disembodied intellect mirrors man's rejection of non-cognitive modes of

experience.

Emshwiller, Carol (1957). "Hunting Machine," *Science Fiction Stories*, May 1957. Reprinted in Dikty (1958). Automated hunting trivializes violence and brutalizes the hunter.

Epernay, Mark, pseud. for J.K. Galbraith (1963). "The Fully Automated Foreign Policy." In: Epernay (1963). Computers replace top government decision-makers.

Epernay, Mark, pseud. for J.K. Galbraith (1963). "The Takeover." In: Epernay (1963). A computerized prediction model is used to gain control of the U.S. Economy.

Forster, E.M. (1928). "The Machine Stops." In E.M. Forster. *The Eternal Moment and Other Stories*. New York: Harcourt Brace Jovanovich, 1928. Also reprinted in Lewis (1963). Automated utopia saps human vitality, and civilization collapses for want of essential skills.

Gold, Herbert (1961). "The Day They Got Boston." Reprinted in Conklin (1963). Faulty rubber band on a deck of punched cards leads to inadvertent bombing of Boston by the Soviets.

Gotlieb, Phyllis (1970). "Score/Score." Reprinted in Clarkson (1970). Teaching machines simulate pupils to keep other teaching machines in order after a decline in the human population.

Goulart, Ron (1964). "Into the Shop." Reprinted in Amis and Conquest (1965), Goulart (1971a). Robot justice proves unreliable.

Goulart, Ron (1965). "Badinage." Reprinted in Nolan and Van Vogt (1965). The ultimate consumer society has a debtor's prison.

Goulart, Ron (1965). "Calling Dr. Clockwork." Reprinted in Wollheim and Carr (1966), Goulart (1971b). Androids operate automated hospital which swallows up human patients.

Goulart, Ron (1969). "Broke Down Engine." Reprinted in Goulart (1971b). Computer simulation of individual life-histories is used to control the human population— projected social utility determines one's fate.

Goulart, Ron (1969). "Lofthouse." Reprinted in Goulart (1971b). Computerized house— replete with magical powers— runs off with owner's girlfriend.

Goulart, Ron (1970). "What's Become of Screwloose." Reprinted in Goulart (1971a). Screwloose is an android who survives his maker but is ultimately destroyed in carrying out his duty.

Goulart, Ron (1971). "Hardcastle." In: Goulart (1971a). Wife forms liaison with computerized house.

Goulart, Ron (1971). "Monte Cristo Complex." In: Goulart (1971a). "Holmes" is a man, "Watson" an android, in this parody of a detective story on the Monte Cristo theme.

Hickey, H.B. (1952). "Hilda." Reprinted in Asimov and Conklin (1963). Robot "loves" man to death, literally.

Inglis, James (1965). "Night Watch." In Carnell (1965). Asov, a cybernetic device, fades in and out of consciousness as it records the decay of the galaxy.

Kagan, Norman (1964). "The Mathenauts," *If*, July 1964. Reprinted in Merril (1965). A lighthearted mathematical excursion into abstract space.

Kagan, Norman (1964). "Four Brands of Impossible," *Fantasy and Science Fiction*, Sept. 1964. Reprinted in Wollheim and Carr (1965).

Kawin, Bruce (1967). "FORM 5640A: Report of a Malfunction," *Data Processing Magazine* 9 (April 1967). Welfare caseworker is replaced by a computer terminal which then breaks down.

Kazantsev, Alexander (1962). "The Martian." In Dutt (1962). Martian manuscript is deciphered by computer.

Knight, Damon (1955). "Dulcie and Decorum," *Galaxy Science Fiction*, March 1955.

Knight, Damon (1956). "Stranger Station." Reprinted in Amis and Conquest (1965). Man-machine affinity emerges in the presence of something truly alien.

Kuttner, Henry (1951). "Those Among Us." Reprinted in Nolan and Van Vogt (1965). The shock of discovering itself to be non-human results in android's self-destruction.

Kuttner, Henry; Moore, C.L. (1955). "Two Handed Engine." Reprinted in Knight (1968). Machine assumes the role of judge and executioner.

LaFarge, Oliver (1951). "John the Revelator," *Fantasy and Science Fiction*, Feb. 1951. Reprinted in Boucher and McComas (1952). This unusual story was written when computers were still a laboratory phenomenon and used largely for military purposes. John is a highly advanced computer— successor to Luke and the earlier Marks— built for the U.S. Department of Defense. The computer exhibits a moral sense, sufficiently embarrassing to provoke remedial action in the form of "lobotomy."

Lafferty, R.A. (1961). "All the People." Reprinted in Lafferty (1970). Conscious robot discovers its identity and allies itself with aliens against its human masters.

Lafferty, R.A. (1964). "What's the Name of that Town." Reprinted in Lafferty (1970). In this parody of scientific research, a computer named Epiktistes comes up with unacceptable results.

Laguna, J. Anthony (1959). "The Predicting Machine," *Computers and Automation* **8** (May 1959). The computer's prediction for the year 2200 is misinterpreted by an unscrupulous individual.

Lamport, Felicia (1961). "A Sigh for Cybernetics," *Harper's Magazine*, Jan. 1961. Reprinted in Merril (1961).

Lawson, Jack B. (1964). "The Competitors," *If*, Jan. 1964. Reprinted in Wollheim and Carr (1965).

Lehman, W.P. (1966). "Decoding the Martian Language," *Data Processing Magazine* **8** (April 1966). A computer is used to process fragments of Martian speech.

Leiber, Fritz (1953). "A Bad Day for Sales." Reprinted in Asimov and Conklin (1963), Silverberg (1968). A mobile sales robot continues its insipid routine as Manhattan disappears in a cloud of radioactive dust.

Leiber, Fritz (1962). "The 64-square Madhouse," *If*, May 1962. Reprinted in Pohl (1966). Computer plays grand-master level chess.

Lem, Stanislaw (1970). "The Twenty-Fourth Journey of Ion Tichy." In Suvin (1970). In this satire on capitalist society, a Governing Machine is designed to assume decision-making functions— with disastrous consequences.

Lem, Stanislaw (1970). "The Computer that Fought a Dragon. " In Suvin (1970). A computer plays the role of Sorcerer's Apprentice with the inadvertent creation of a cyberdragon.

Leman, Grahme (1972). "Conversational Mode." Reprinted in Pohl (1972). Computer as psychotherapist.

Lowenkopf, Shelly (1965). "The Addict." Reprinted in Nolan and Van Vogt (1965). Neurotic female android takes revenge on male android for "sexual advances."

McIntosh, J.T. (1951). "Machine Made." Reprinted in Conklin (1965). An evolving computer induces a moronic cleaning woman to build a contraption which makes her more intelligent.

McIntosh, J.T. (1963). "Spanner in the Works," *Analog Science Fact and Science Fiction*, March 1963. Genius, a multi-million

dollar computer which helps run Terra's Intelligence Department, must be "debriefed" as a result of sabotage by a former Department Head.
Mayfield, M.I. (1959). "On Handling the Data," *Astounding Science Fiction*, Sept. 1959.
Miller, Walter M. Jr. (1954). "I Made You." Reprinted in Knight (1968). An automated killing machine destroys its maker.
Miller, Walter M. Jr. (1952). "Dumb Waiter," *Astounding Science Fiction*, April 1952. Reprinted in Conklin (1954). An abandoned, automated city is saved by ingenious tampering with the central computer.
Niven, Larry (1965). "Becalmed in Hell." Reprinted in Knight (1966), Ferman and Mills (1970). An apparent malfunction in a cyborg ship— thought to be psychosomatic— turns out to be a mechanical problem.
Nolan, William F. (1954). "The Joy of Living." Reprinted in Nolan and Van Vogt (1965). A man discovers human feelings in his robot "wife" and decides not to return her to the factory.
Okhotnikov, Vadim (1963). "The Fiction Machines." Reprinted in Magidoff (1963). A scientist-engineer develops a series of machines to enable himself to become a writer, but eventually has to resort to conventional methods.
Opler, Ascher (1966). "Bon Voyage— 1984 Style," *Datamation* **12** (Jan. 1966). Reprinted in Moshman (1966). A benevolent computer handles all the details of a man's vacation plans.
Phelan, R.C. (1960). "Something Invented Me." Reprinted in Merril (1961). Man's use of a computerized writing machine leads to identity crisis.
Piper, H. Beam (1958). "Graveyard of Dreams," *Galaxy Science Fiction*, Feb. 1958. Disillusioned student of computer theory is faced with the problem of informing his compatriots that the supercomputer of their dreams is just a myth.
Pohl, Frederik (1954). "The Midas Plague." Reprinted in Pohl (1957). Robots turn out to be more reliable than people in meeting consumption quotas.
Pohl, Frederik (1966). "Day Million." Reprinted in Wollheim and Carr (1967). Ten thousand years hence the pleasures of the flesh are replaced by electronically mediated simulacra.
Pohl, Frederik (1968). "Schematic Man." Reprinted in Pohl (1970). Computer expert's attempt to teach a computer to duplicate the former's behavior leads to loss of identity.

Powers, William T. (1953). "Allegory," *Astounding Science Fiction*, April 1953. Reprinted in Conklin (1960). Research funds are peremptorily denied to an ingenious inventor because the computers reject his discoveries as contrary to known fact.

Price, George R. (1957). "The Maker of Computing Machines," *Computers and Automation* 6 (Oct. 1957). Human History is represented in the form of a parable of computer evolution.

Rackham, John (1966). "Computer's Mate." In Carnell (1966b). Man with congenital brain defect is complemented in a symbiotic relationship with a shipboard computer. The defect proves to be a vital asset when his party of space travellers encounters an alien intelligence.

Rankine, John (1966). "Six Cubed Plus One." In Carnell (1966a). Chance orientation of a group of teaching machines results in the formation of a powerful, conscious entity.

Rice, Elmer L. (1922). "The Adding Machine." Reprinted in Mersand (1964). Pre-computer dramatization of the meaninglessness of life in industrialized society.

Rolland, Romain (1932). *The Revolt of the Machines or Inventions Run Wild, A Motion Picture Fantasy* (trans. from the French by William A. Drake). Ithaca, N.Y.: The Dragon Press. Machines run amok in this pre-computer play about man's abiding obtuseness.

Ryan, James F. (1966). "Maximillian the Great," *Data Processing Magazine* 8 (Nov. 1966). A domestic computer creates friction in the household.

Saberhagen, Fred (1962). "Without a Thought." Reprinted in Silverberg (1968). Man's animal cunning triumphs over vastly superior destructive machines.

Saberhagen, Fred (1965). "Masque of the Red Shift." Reprinted in Wollheim and Carr (1966), Pohl (1968). Brutal machines clash with brutal human beings.

Salomon, Louis B. (1958). "Univac to Univac *(sotto voce)*," *Harper's Magazine*, March 1958. Reprinted in Baker (1963). Two computers speculate about human beings.

Sandberg, Richard T. (1967). "The Perfect Crime," *Data Processing Magazine* 9 (Feb. 1967). Computerized mate selection works all too well by bringing together two people with complementary, murderous schemes.

Schenk, Hilbert Jr. (1959). "Me." Reprinted in Fadiman (1962).

Shaara, Michael (1956). "2066: Election Day," *Astounding Science Fiction*, Dec. 1956. Reprinted in Dikty (1958). A crisis develops as computerized testing for president fails to unearth a qualified candidate.

Sheckley, Robert (1953). "Ask a Foolish Question," *Science Fiction Stories*, no. 53, 1953.

Sheckley, Robert (1953). "Fool's Mate." Reprinted in Knight (1968). The reassertion of human will plucks victory from defeat in a computerized space war.

Sheckley, Robert (1968). "Street of Dreams, Feet of Clay." Reprinted in Wollheim and Carr (1969). The urban jungle proves preferable to life in a computer-controlled, model city.

Sheckley, Robert (1969). "Can You Feel Anything When I Do This?" Reprinted in Playboy (1971b). Housewife reacts violently to impulses aroused in herself by an amorous robot.

Silverberg, Robert (1957). "Run of the Mill," *Astounding Science Fiction*, July 1957. Faith in computerized forecasting sets up the conditions for self-fulfilling prophecy.

Silverberg, Robert (1956, revised 1968). "The Macauley Circuit." Reprinted in Silverberg (1968). The synthesis and composition of music by machine renders the creative musician obsolete.

Silverberg, Robert (1968). "Going Down Smooth." Reprinted in Wollheim and Carr (1969). A psychiatric computer exhibits a capacity for neurosis and deception.

Simak, Clifford D. (1944). "Huddling Place." Reprinted in Silverberg (1970b). The habit of isolation— despite or perhaps because of instantaneous communications— constrains man's ability to act.

Simpson, Alan (1961). "The Twenty-Third Psalm— Modern Style." In: Alan Simpson, Liberal education in a university, *Washington University Magazine*, Feb. 1961. An extended version appears in Baker (1963).

Sladek, John T. (1967). "The Happy Breed." In Ellison (1967). Under centralized, computer control human beings regress to a utcrine state in a world without pain.

Slesar, Henry (1958). "Examination Day." Reprinted in Playboy (1971b). Computerized intelligence-testing by government serves ruthlessly to preserve the status quo.

Solzhenitsyn, Aleksandr I. (1973). *Candle in the Wind* (trans. from the Russian by Keith Armes). Minneapolis, Minn.: University of Minnesota Press. The computer is but a manifestation of scientific rationality in this drama of conflict between amoral progress and humanistic values.

Strauss, Richard H. (1965). "I Think, Therefore," in: *Perspectives in Biology and Medicine*. Chicago: University of Chicago Press, Summer 1965. Reprinted in *Data Processing Magazine* 8 (July 1966). This story is narrated by a computer that has inherited the identity of a man.

Strugatsky, Arkady and Boris (1962). "Spontaneous Reflex." Reprinted in Dutt (1962). A self-programming robot acts in unanticipated fashion and must be stopped by force.

Sturgeon, Theodore (1965). "The Nail and the Oracle." Reprinted in Playboy (1971b). Three scheming individuals are thwarted by an advanced military computer that heeds the "THINK" admonition.

Szilard, Leo (1961). "Calling All Stars." In Szilard (1961); reprinted in Baker (1963). Intelligent computer-entity warns of ominous developments on Earth.

Townes, Robert Sherman (1952). "Problem for Emmy," *Startling Stories*, June 1952. Reprinted in Conklin (1954), Lewis (1963). Emmy is a huge computer that shows signs of a burgeoning self-awareness.

Van Scyoc, Sydney (1968). "A Visit to Cleveland General." Reprinted in Wollheim and Carr (1969). Automated medical care in the future is shown as a nightmare of arbitrary decisions and disregard for elementary human rights.

Van Vogt, A.E. (1951). "Fulfillment." Reprinted in Healy (1951). A time-tripping supercomputer is overcome by the combined forces of its ancestor and that earlier machine's human designer.

Varshavsky, Ilya (1970). "SOMP." In Suvin (1970). In this satire on the computer as intelligence amplifier, a scientist is exposed to ridicule.

Vonnegut, Kurt Jr. (1950). "Epicac." Reprinted in Vonnegut (1968). A military computer burns itself out over an unrequited love and the realization of its machinehood.

Walde, A.A. (1966). "Bircher." Reprinted in Wollheim and Carr (1967). In this bizarre detective story set in a computerized society of the future, a machine-selected police commissioner unravels the mystery of a homocide.

Weiss, Milton (1960). "Computer Bergerac," *Computers and Automation* 9 (May 1960). Computer plays cupid for shy programmer.

Wells, Robert W. (1971). "Alicia Marches on Washington," *Saturday Evening Post*, Jan. 21, 1971. Righteous citizen confronts erring tax computer.

Williamson, Jack (1947). "With Folded Hands." Reprinted in Silverberg (1968). Humanoids protect man from himself to the point of annihilating the human spirit.

Woodside, B. (1966). "The Nano Bandits," *Data Processing Magazine* 8 (Dec. 1966). Computer fraud on a colossal scale.

Wyndham, John (1954). "Compassion Circuit." Reprinted in Conklin (1960), Amis and Conquest (1965). Shocked husband falls down stairs after discovering that his ailing wife has been transformed into a robot.

Zelazny, Roger (1966). "The Keys to December." Reprinted in Wollheim and Carr (1967). A parable of technological hybris, and the unity of intelligent life-forms.

Zelazny, Roger (1966). "For a Breath I Tarry," *New Worlds*, 1966. Reprinted in Allison *et al.*(1973). A computer recreates man after the human race has disappeared from the Earth.

Anthologies

Aldiss, Brian (1959). *No Time Like Tomorrow*. New York: New American Library.
Aldiss, Brian (1965). *Who Can Replace a Man?* New York: Harcourt Brace Jovanovich.
Allison, Leonard; Jenkin, Leonard; Perrault, Robert, eds. (1973). *Survival Printout*. New York: Random House.
Amis, Kingsley; Conquest, Robert, eds. (1961). *Spectrum: A Science Fiction Anthology*. New York: Harcourt Brace Jovanovich.
Amis, Kingsley; Conquest, Robert, eds. (1965). *Spectrum 4*. New York: Harcourt Brace Jovanovich.
Anderson, Poul (1969). *Seven Conquests*. New York: Macmillan.
Asimov, Isaac (1950). *I, Robot*. Garden City, N.Y.: Doubleday.
Asimov, Isaac (1957). *Earth Is Room Enough*. Garden City, N.Y.: Doubleday.
Asimov, Isaac (1959). *Nine Tomorrows*. Garden City, N.Y.: Doubleday.
Asimov, Isaac (1964). *The Rest of the Robots*. Garden City, N.Y.: Doubleday.
Asimov, Isaac; Conklin, Groff, eds. (1963). *Fifty Short Science Fiction Tales*. New York: Collier Books.
Baker, Robert A., ed. (1963). *A Stress Analysis of a Strapless Evening Gown*. Englewood Cliffs, N.J.: Prentice-Hall.
Ballard, J.G. (1963). *The Four-Dimensional Nightmare*. London: Victor Gollancz.
Barthelme, Donald (1968). *Unspeakable Practices, Unnatural Acts*. New York: Farrar, Straus and Giroux.
Borges, Jorge Luis (1962). *Labyrinths* (selected stories and other writings edited by Donald A. Yates and James E. Irby). New York: New Directions.

ANTHOLOGIES

Boucher, Anthony; McComas, J. Francis, eds. (1952). *The Best from Fantasy and Science Fiction*. Boston: Little, Brown and Company.
Boulle, Pierre (1966). *Time Out of Mind* (trans. from the French by Xan Fielding and Elisabeth Abbott). New York: Vanguard Press.
Bova, Ben (1973). *Forward in Time*. New York: Popular Library.
Bradbury, Ray (1969). *I Sing the Body Electric*. New York: Alfred A. Knopf.
Carnell, John, ed. (1965). *New Writings in SF-3*. London: Dennis Dobson.
Carnell, John, ed. (1966a). *New Writings in SF-7*. London: Dennis Dobson.
Carnell, John, ed. (1966b). *New Writings in SF-8*. London: Dennis Dobson.
Clarke, Arthur C. (1967). *The Nine Billion Names of God: The Best Short Stories of Arthur C. Clarke*. New York: Harcourt Brace Jovanovich.
Clarkson, Stephen, ed. (1970). *Visions 2020*. Edmonton, Alberta: M.G. Hurtig, Ltd.
Conklin, Groff, ed. (1954). *Science-Fiction Thinking Machines*. New York: Vanguard Press.
Conklin, Groff, ed. (1960). *13 Great Stories of Science Fiction*. Greenwich, Conn.: Fawcett Publications.
Conklin, Groff, ed. (1963). *17 x Infinity*. New York: Dell Publishing Co.
Conklin, Groff, ed. (1965). *Giants Unleashed*. New York: Grosset and Dunlop.
Dick, Philip K. (1957). *The Variable Man and Other Stories*. New York: Ace Books.
Dick, Philip K. (1969). *The Preserving Machine*. New York: Ace Books.
Dikty, T.E., ed. (1958). *The Best Science-Fiction Stories and Novels: Ninth Series*. Chicago: Advent Publishers.
Disch, Thomas M., ed. (1971). *The Ruins of the Earth*. New York: G.P. Putnam.
Dutt, Violet L., trans. (1962). *Soviet Science Fiction*. New York: Collier Books.
Ellison, Harlan (1967). *I Have No Mouth and I Must Scream*. New York: Pyramid Books.

Ellison, Harlan, ed. (1967). *Dangerous Visions*. New York: New American Library.
Epernay, Mark, pseud. for John Kenneth Galbraith (1963). *The McLandress Dimension*. Boston: Houghton Mifflin.
Fadiman, Clifton, ed. (1958). *Fantasia Mathematica*. New York: Simon and Schuster.
Fadiman, Clifton, ed. (1962). *The Mathematical Magpie*. New York: Simon and Schuster.
Ferman, Edward L.; Mills, Robert P., eds. (1970). *Twenty Years of the Magazine of Fantasy and Science Fiction*. New York: G.P. Putnam's Sons.
Goulart, Ron (1971a). *What's Become of Screwloose? And Other Enquiries*. New York: Scribner's Sons.
Goulart, Ron (1971b). *Broke Down Engine*. New York: Macmillan.
Healy, Raymond J., ed. (1951). *New Tales of Space and Time*. New York: Pocket Books.
Knight, Damon, ed. (1964). *A Century of Great Short Science Fiction Novels*. New York: Dell Publishing Co.
Knight, Damon, ed. (1966). *Nebula Award Stories, 1965*. Garden City, N.Y.: Doubleday.
Knight, Damon, ed. (1967). *Orbit 2*. New York: G.P. Putnam's Sons.
Knight, Damon, ed. (1968a). *The Metal Smile*. New York: Belmont Tower Books.
Knight, Damon, ed. (1968b). *One Hundred Years of Science Fiction*. New York: Simon and Schuster.
Knight, Damon, ed. (1968c). *Orbit 3*. New York: Berkley Publishing Co.
Knight, Damon, ed. (1972). *Orbit 11*. New York: G.P. Putnam's Sons.
Lafferty, R.A. (1970). *Nine Hundred Grandmothers*. New York: Ace Publishing Co.
Lem, Stanislaw (1974). *The Cyberiad: Fables for the Cybernetic Age* (trans. from the Polish by Michael Kandel). New York: Seabury Press.
Lewis, Arthur O. Jr., ed. (1963). *Of Men and Machines*. New York: E.P. Dutton.
Magidoff, Robert, ed. (1963). *Russian Science Fiction*. New York: New York University Press.
Merril, Judith, ed. (1961). *The Year's Best S-F*. New York: Simon and Schuster.
Merril, Judith, ed. (1965). *The Year's Best S-F: 10th Annual Edition*.

ANTHOLOGIES

New York: Delacorte Press.
Mersand, Joseph, ed. (1964). *Three Plays About Business in America*. New York: Simon and Schuster.
Moshman, Jack, ed. (1966). *Faith, Hope and Parity*. Washington, D.C.: Thompson Book Co.
Moskowitz, Samuel, ed. (1963). *The Coming of the Robots*. New York: Collier Books.
Nolan, William F., ed. (1965). *The Pseudo-People: Androids in Science Fiction*. Los Angeles: Sherbourne Press.
Nolan, William F., ed. (1970). *The Future is Now*. Los Angeles: Sherbourne Press. Contains a bibliography on robot fiction.
Playboy Science Fiction (1971a). *Last Train to Limbo*. Chicago: Playboy Press.
Playboy Science Fiction (1971b). *From The "S" File*. Chicago: Playboy Press.
Playboy Science Fiction (1971c). *The Fully Automated Love Life of Henry Keanridge*. Chicago: Playboy Press.
Pohl, Frederik, ed. (1954). *Assignment in Tomorrow*. Garden City, N.Y.: Doubleday.
Pohl, Frederik, ed. (1965). *The Case Against Tomorrow: Science Fiction Short Stories*. New York: Ballantine Books.
Pohl, Frederik, ed. (1966). *The If Reader of Science Fiction*. Garden City, N.Y.: Doubleday.
Pohl, Frederik, ed. (1968). *The Second IF Reader of Science Fiction*. Garden City, N.Y.: Doubleday.
Pohl, Frederik (1970). *Day Million*. New York: Ballantine Books.
Pohl, Frederik, ed. (1972). *Best Science Fiction for 1972*. New York: Ace Books.
Prokofieva, R., trans. (1962). *More Soviet Science Fiction*. New York: Collier Books.
Quinn, James L.; Wulff, Eve, eds. (1957). *The First World of If*. Kingston, N.Y.: Quinn Publishing Co.
Quinn, James L.; Wulff, Eve, eds. (1958). *The Second World of If*. Kingston, N.Y.: Quinn Publishing Co.
Silverberg, Robert, ed. (1968). *Men and Machines*. New York: Meredith Press.
Silverberg, Robert, ed. (1970a). *The Mirror of Infinity*. New York: Harper and Row.
Silverberg, Robert, ed. (1970b). *Science Fiction Hall of Fame Vol. I*. Garden City, N.Y.: Doubleday.
Suvin Darko, ed. (1970). *Other Worlds, Other Seas*. New York: Random House.

Szilard, Leo (1961). *The Voice of the Dolphins*. New York: Simon and Schuster.
Vonnegut, Kurt (1968). *Welcome to the Monkey House*. New York: Delacorte Press.
Wollheim, D.A.; Carr, T., eds. (1965). *World's Best Science Fiction: 1965*. New York: Ace Books.
Wollheim, D.A.; Carr, T., eds. (1966). *World's Best Science Fiction: 1966*. New York: Ace Books.
Wollheim, D.A.; Carr, T., eds. (1967). *World's Best Science Fiction: 1967*. New York: Ace Books.
Wollheim, D.A.; Carr, T., eds. (1968). *World's Best Science Fiction: 1968*. New York: Ace Books.
Wollheim, D.A.; Carr, T., eds. (1969). *World's Best Science Fiction: 1969*. New York: Ace Books.
Wollheim, D.A.; Carr, T., eds. (1971). *World's Best Science Fiction: 1971*. New York: Ace Books.

Criticism and Other Non-Fiction

Aldiss, Brian (1973). *Billion Year Spree*. Garden City, N.Y.: Doubleday.

Amis, Kingsley (1960). *New Maps of Hell*. New York: Harcourt Brace Jovanovich.

Ascher, Marcia (1962). Fictional computers and their themes, *Computers and Automation*, Dec. 1962.

Ascher, Marcia (1963). Computers in science fiction, *Harvard Business Review* **41**, pp. 40-45. Reprinted in Taviss (1970).

Ascher, Marcia (1973). Computers in science fiction— II, *Computers and Automation*, Nov. 1973.

Ash, Brian (1975). *Faces of the Future*. New York: Taplinger Publishing Co.

Asimov, Isaac (1973). *Today and Tomorrow and ....* Garden City, N.Y.: Doubleday.

Baer, Robert M. (1972). *The Digital Villain*. Reading, Mass.: Addison-Wesley. Contains a lengthy discussion of selected works of computer-related fiction.

Bailey, James Osler (1947). *Pilgrims Through Space and Time*. New York: Argus Books.

Bellow, Saul, *et al.*(1975). *Technology and the Frontiers of Knowledge*. Garden City, N.Y.: Doubleday.

Berneri, Marie Louise (1951). *Journey Through Utopia*. London: Routledge and Kegan Paul.

Block, Henry; Ginsburg, Herbert (1968). The psychology of robots, *Psychology Today*, April 1968. Reprinted in Pylyshyn (1970).

Boguslav, Robert (1965). *The New Utopians: A Study of System Design and Social Change*. Englewood Cliffs, N.J.: Prentice-Hall.

Bretnor, Reginald, ed. (1974). *Science Fiction, Today and Tomorrow*. New York: Harper and Row.

Clareson, Thomas D., ed. (1971). *The Other Side of Realism*. Bowling Green, Ohio: Bowling Green University Popular Press. A collection of critical papers on science fiction.
Clareson, Thomas D. (1972). *Science Fiction Criticism: An Annotated Checklist*. Kent, Ohio: Kent State University Press. Contains an annotated bibliography of critical writings on science fiction.
Cohen, John (1966). *Human Robots in Myth and Science*. London: George Allen and Unwin.
Culbertson, James T. (1963). *The Minds of Robots*. Urbana, Ill.: University of Illinois Press.
Davenport, Basil, et al.(1964). *The Science Fiction Novel*. Chicago: Advent Publishers.
Franklin, H. Bruce (1965). Science fiction as an index to popular attitudes toward science: a danger, some problems, and two possible solutions, *Extrapolations* 6 (May 1965), pp. 23-31.
Franklin, H. Bruce, ed. (1966). *Future Perfect*. New York: Oxford University Press.
Franklin, H. Bruce (1972). Chic bleak in fantasy fiction, *Saturday Review*, July 15, 1972.
Gattegno, Jean (1971). *La Science-Fiction*. Paris: Presses Universitaires de France.
Ginestier, Paul (1961). *The Poet and the Machine* (trans. from the French by Martin B. Friedman). Chapel Hill: University of North Carolina Press.
Gotlieb, C.C.; Borodin, A. (1973). *Social Issues in Computing*. New York: Academic Press.
Gould, Heywood (1971). *Corporation Freak*. New York: Tower Publications.
Handlin, Oscar (1965). Science and technology in popular culture, *Daedalus* 94 (Winter 1965), pp. 156-170.
Hillegas, Mark (1963). Science fiction as a cultural phenomenon: a re-evaluation, *Extrapolations* 4 (May 1963), pp. 26-33.
Hillegas, Mark (1967). *The Future as Nightmare: H.G. Wells and the Anti-Utopians*. New York: Oxford University Press.
Hurley, Neil P. (1969). Coming of the humanoids, *Commonweal* 5 (Dec. 1969), pp. 297-300.
Huxley, Aldous (1963). *Literature and Science*. New York: Harper and Row.
Lee, Robert S. (1966). The computer's public image, *Datamation* 12 (Dec. 1966). Pp. 33-35.

Lundwall, Sam J. (1971). *Science Fiction: What It's All About*. New York: Ace Books.
Mailer, Norman (1970). *Of a Fire on the Moon*. Boston: Little, Brown.
Marx, Leo (1964). *The Machine in the Garden: Technology and the Pastoral Ideal in America*. New York: Oxford University Press.
Matusow, Harry M. (1968). *The Beast of Business: A Record of Computer Atrocities*. London: Wolfe Publishing, Ltd.
Moore, Patrick (1957). *Science and Fiction*. London: George G. Harrap.
Moskowitz, Samuel (1963). *Explorers of the Infinite*. Cleveland: World Publishing Co.
Moskowitz, Samuel (1966). *Seekers of Tomorrow: Masters of Science Fiction*. Cleveland: World Publishing Co.
Mowshowitz, Abbe (1976). *The Conquest of Will: Information Processing in Human Affairs*. Reading, Mass.: Addison-Wesley.
Mumford, Lewis (1922). *The Story of Utopias*. Reprinted in 1962, Viking Press, New York.
Pilgrim, John (1963). Science fiction and anarchism, *Anarchy* 3 (Dec. 1963), pp. 361-375.
Plank, Robert (1963;1965). The Golem and the robot, *Literature and Psychology*, pp. 13-15; pp. 12-27.
Pylyshyn, Zenon W., ed. (1970). *Perspectives on the Computer Revolution*. Englewood Cliffs, N.J.: Prentice-Hall.
Rose, Lois; Rose, Stephen (1970). *The Shattered Ring*. Richmond, Va.: John Knox Press.
Sadoul, Jacques (1973). *Histoire de la Science-Fiction Moderne*. Paris: A. Michel.
Sieman, Frederick (1971). *Science Fiction Story Index*. Chicago: American Library Association.
Sussman, Herbert L. (1968). *Victorians and the Machine: The Literary Response to Technology*. Cambridge, Mass.: Harvard University Press.
Sypher, Wylie (1968). *Literature and Technology*. New York: Vintage.
Taviss, Irene, ed. (1970). *The Computer Impact*. Englewood Cliffs, N.J.: Prentice-Hall.
Taylor, Gordon Rattray (1963). The age of the androids, *Encounter*, Nov. 1963.

Tuck, Donald Henry, compiler (1974). *The Encyclopedia of Science Fiction and Fantasy Through 1968*. Chicago: Advent.
Van Tassel, Dennie L. (1976). *The Compleat Computer*. Chicago: Science Research Associates. A miscellany of fiction and non-fiction dealing with computers.
Weizenbaum, Joseph (1976). *Computer Power and Human Reason: From Judgment to Calculation*. San Francisco: W.H. Freeman.
Wollheim, Donald A. (1971). *The Universe Makers*. New York: Harper and Row.

Index

Commentary Index: Names

Alban, Antony, 2, 3, 85, 195, 229
Aldiss, Brian, 83, 230, 267
Alfvén, Hannes
  see Johannesson, Olof
Anderson, Poul, 2, 85, 266
Asimov, Isaac, xvi, 3, 39, 40, 84, 120, 168, 230, 266

Bacon, Francis, 39, 166
Balchin, Nigel, 227
Ballard, J.G., 4, 230
Banks, Raymond E., 86
Barth, John, 1, 4, 43
Barthelme, Donald, 1
Bellamy, Edward, 81
Benford, Gregory, 122
Bierce, Ambrose, 168
Blish, James, 3, 44, 83, 227, 264
Boulle, Pierre, 227
Bova, Ben, 4, 166
Bradbury, Ray, 229, 264
Brown, Frederic, 40
Brunner, John, 3, 43, 120, 121, 195, 196
Buck, Doris P., 168
Burdick, Eugene, 3, 120, 197
Butler, Samuel, 266

Caidin, Martin, 2, 41, 42, 194
Cameron, Lou, 2, 122
Capek, Karel, 231
Carlson, Mel, 166
Clarke, Arthur C., 2, 4, 39, 40, 42, 43, 85, 121, 230, 267

Clement, Hal, 4
Cole, Burt, 2, 82, 84
Compton, David G., 2, 120, 121, 197, 230
Conklin, Groff, xvi
Corwin, Norman, 4

del Rey, Lester, 231, 266
Delany, Samuel R., 264
Dick, Philip K., 3, 119
Dickson, Gordon, 168
Dnieprov, Anatoly, 4, 41
Doerr, Edd, 167
Dozois, Gardner R., 3, 195
Draper, Hal, 166

Eklund, Gordon, 196
Ellison, Harlan, 41, 169, 266
Ellul, Jacques, 43
Emschwiller, Carol, 196
Epernay, Mark, 3, 119
Escarpit, Robert, 4

Fairman, Paul, 82, 84, 85, 194, 229, 230
Forster, E.M., 85, 229

Galbraith, John Kenneth
  see Epernay, Mark
Gold, Herbert, 167
Gotlieb, Phyllis, 3
Goulart, Ron, 122, 123, 228

Heinlein, Robert, 1, 2, 44, 264

335

Hickey, H.B., 266
Hodder-Williams, Christopher, 2, 43, 197
Huxley, Aldous, 82, 84, 85, 193

Johannesson, Olof, 83, 267
Jones, D.F., 42, 265

Kawin, Bruce, 165
Knight, Damon, 265
Kuttner, Henry, 122, 265

Lafferty, R.A., 4, 44, 265
Le Guin, Ursula, 264
Lehmann, W.P., 4
Leiber, Fritz, 123
Lem, Stanislaw, 41
Levin, Ira, 2, 82, 84, 85, 230

McIntosh, J.T., 231
Maine, Charles Eric, 2, 42, 194
Miller, Walter M., Jr., 121
Moorcock, Michael, 267
Moore, C.L., 122
Morris, William, 81

Nolan, William F., 266

Orwell, George, 82, 84, 85, 193

Phelan, R.C., 4
Pohl, Frederik, 123
Powers, William T., 84
Price, George R., 263

Rankine, John, 267
Rice, Elmer L., 263
Ryan, James F., 167

Saberhagen, Fred, 228
Shaara, Michael, 3, 119
Sheckley, Robert, 4, 86, 121, 167, 196, 228
Shelley, Mary, 168
Silverberg, Robert, 4, 266

Sladek, John, 2, 3, 4, 195, 196
Slesar, Henry, 84
Solzhenitsyn, Aleksandr I., 2, 4, 165
Strugatsky, Arkady and Boris, 41
Sturgeon, Theodore, 4, 122

Townes, Robert S., 265
Trimble, Louis, 83
Tyler, Theodore, 39, 167

Villiers de l'Isle-Adam, 265
Vonnegut, Kurt, Jr., 1, 3, 44, 82, 168

Wells, H.G., 81-82, 83
Woods, William C., 121

Zamiatin, Eugene, 82, 84
Zelazny, Roger, 231

Commentary Index: Subjects

alienation, 194
  see also responsibility
androids, xvi
animal cunning, 227-228
anthropomorphism, 228
anti-utopia, xvii, 82
  see also dystopia
asexual mechanism, 228
atrophy of
  human capabilities, 228
  judgment, 121
automated decision-making, 41, 118
automated hunting, 196
automated society
  creativity, 230
  in fiction, 81
  innovation, 85
automation
  cost, xiv
  dehumanization, 165
  diminishing role of humans, xvi
  displacement of humans, xvii
  impact on social institutions, xvii, 4
  machine failure, 166
  malaise, 168
  vulnerability of society, xix
autonomous machines, 168, 267
autonomous technology, 43, 123

behavioral monitoring, 3, 82, 193, 195-196
benevolent kingship, 83

bureaucratization, 4

capacity for neurosis, 266
Census, Bureau of, 164
centralization of power, 44, 81
centralized services, 84
character types, 1
citizen participation
  see participatory democracy
citizenship, responsibility of, 120
  see also responsibility
cognitive dissonance, 44
collective suicide, 86
complexity of behavior, xx
complexity of government, 120
computer applications
  accounting, 168
  corporate management, xiii, 3, 119
  diffusion of, xiii, 164
  domestic, 167
  economic modeling, 3
  education, xiii, xvii, 3, 195
  health-care, xiii, xvii, 3, 119, 195
    see also behavioral monitoring
  military, 4, 42, 85, 119, 121
    see also war gaming
  opinion polling, 2-3, 120
  scheduling, 167
  scientific research, 3
  skills testing, 3
  welfare, 165-166
computer as
  bureaucrat, 84

337

computer as *continued*
   demon, 194
   God, xx, 39-40
   Golem, 39
   hypnotist, 194
   telepathist, 194
computer-assisted diagnosis
   see computer applications, health-care
computer-assisted instruction
   see computer applications, education
computer-communications, xiii, 165, 193, 197-198
   see also computer network
computer controlled community, 86
computer errors, 167
computer-fortress, xx, 41
computer network, xiv, 42, 43
   see also computer-communications
computer perversion, 266
computer-related fiction
   definition of, xv-xvi
   themes, xvi-xxi
   types, xvi
computer simulation, 44
   see also simulation of voting behavior
computer systems
   programming errors, xviii, 121
   users' lament, xvii-xviii, 167-168
   vulnerability of, 121
computer terminal, 165
computer utilities, xiv
   see also computer-communications; *and* computer network
computer worship, 196-197
computerized decision-making, xiii, xvii, 168
computerized elections, 120
   see also computer applications, opinion polling
computerized poety, 230
computers
   attitudes toward, 5, 165
   conscious malevolence, xix, 122
      see also machine consciousness
   fictional themes, 2-5
   impact on values, xiv
   over-reliance on, 4
      see also dependence on machines
computers and controlled violence, 196
   see also social engineering
computers and identity, 5
computers and social power, 43
computers in education
   see computer applications, education
computers in everday life, 4
computers in general administration, 194
computers in government, xiii, 119
computers in medicine
   see computer applications, health-care
   and behavioral monitoring
computers in social services, 194-195
computers in the arts, xvii, 4
conscious machines
   see computers, conscious malevolence
   and machine consciousness
consumer society, 122-123
control technology, 83
cosmic mind, xvii, xx, 40, 267
cyborg, xvi

databanks, xiv
   see also privacy
democratic pluralism, 44
dependence on machines, xix, 197-198, 229
   see also computers, over-reliance on

deviation from social norms, 82
distribution of computing power, 85
dossier system, 194
  see also behavioral monitoring; privacy; and surveillance
dystopia, 82, 84, 85
  see also utopia

Eckert-Mauchly Computer Corporation, 164
economic organization, 83
efficiency, xiii, xvii, 81, 118
Electronic Funds Transfer Systems (EFTS), xiv
electronic paradise, xvii
elites, 83
emergence of new life-form, 267
escape from automated society, 195-196

factory discipline, 81
factory system, 118
facts and judgment, 44
forbidden knowledge, 39-40
  see also ineffable names and power of knowledge
free will, 43

Golem, 41-43
  see also computer as Golem
government computer applications
  see computers in government

hero as savior, 86
heroism, 194
home terminal, 165
human choice, 228
human identity, xiv, xix, 226-228, 263-264
  see also quest for identity
human superfluousness, xviii
human superiority, 230
human uniqueness, xiv, xviii
human vs. machine leadership, 119

humane computing, 167
humanoids, xvi

idea of nature, 262
identity
  see human identity and quest for identity
immortality, longing for, 229
individuality as neurosis, 195
industrial army, 81
industrial revolution, 82, 118, 229, 264
industrial society, xvii, 5
industrial technology, 263
ineffable names, 39-40
  see also forbidden knowledge and power of knowledge
information and organization, 39
information explosion, 166
information-processing concepts, 226
information retrieval, 120
information technology, diffusion of, 5
  see also computer applications, diffusion of
information transactions
  see record-keeping
intelligence-testing
  see skills testing
intelligent machines
  see machine intelligence

justice, administration of, 122

kakatopia, 84
  see also dystopia
knowledge
  direct vs. indirect, 166
  key to secrets, xx

large-scale systems, xiii, xix, 167
Luddite action, 168

machine betrayal, 168
  see also computers, conscious malevolence
machine consciousness, xix-xx, 42, 168, 194, 264-265, 267
  see also computers, conscious malevolence
machine intelligence, xiv, xvii-xix, 226-227, 265-266
malaise, 85
  see also alienation
man as machine, 228
man-machine accomodation, xx, 262
man-machine affection, 265-266
man-machine comparisons
  capacity to err, 227
  emotionality, 227
  flexibility, 121
  human uniqueness, xiv
    see also human uniqueness
  information-processors, 40
  versatility, 227
man-machine interaction, xvii, xx, 165
man-machine partnership, xx, 264-265
man-machine relationships, 263-264
man-machine symbiosis, 267
management information systems
  see computer applications, corporate management; and computerized decision-making
master-slave relationships, xx, 266-267
mechanistic culture, 229
mechanistic models, 226-227
mechanization of life, 168-169
medical computer applications
  see computer applications, health-care
Melander and Corydon, myth of, 230
meritocracy, 81
military computers
  see computer applications, military; and war-gaming
mind-altering drugs, 193, 195
mini-computers, 85

organism and mechanism, 263

participation in decision-making, 197
participatory democracy, xiv, 84-85
personal responsibility
  see responsibility
police computers, 85
  see also behavioral monitoring; privacy; and surveillance
post-industrial society, 83
power of knowledge, 39-41
  see also forbidden knowledge and ineffable names
powerlessness, 122
pre-scientific thought, xviii
predicting the future, xv, xix, 262-263
privacy, xiv, 82, 84
  see also databanks; behavioral monitoring; and surveillance
prodigal computers, 166-167
productive efficiency
  see efficiency
psychiatric computer, 266-267
psychotherapy, 195
pursuit of knowledge, 39, 166
  see also knwledge, direct vs. indirect

quest for identity, xviii, 265
  see also human identity

rationalization of production, 118
  see also efficiency
rebel in utopia, 229-230
record-keeping, xiii, 4, 82, 83-84
regulation of computer networks, 198
rejection of direct experience, 229
  see also knowledge, direct vs. indirect
remote terminal, xiii, 165
responsibility, xiv, 43, 120, 122, 194, 230-231

COMMENTARY INDEX: SUBJECTS 341

resurrection of man, xix, 230-231
robots, xvi

scientific progress, 165
self-justifying tools, 166
simulation
 see computer simulation
simulation of voting behavior, 197
 see also computer applications,
  opinion polling
Sisyphus, myth of, 230
skills testing, 119
social complexity, 83
social cybernetics, 165
social engineering, 120, 193, 197
social expression, 122
social management, 84
social organism, 193
social planning, 84
socialist reformers, 81
sociopathic behavior, 122
Sorcerer's Apprentice, xix, 41, 43
supercomputer, 168, 265, 267
surveillance, 4, 82, 84, 120, 193-194
 see also behavioral monitoring;
  and privacy
surveillance devices, 193

teaching machines, 267
 see also computer applications,
  education
technique, 43
technological metaphors, 226
technology, control and use, 2
technology and culture, xvi, 226-227
technology and human imperfection,
 165
thinking machines, xiv
 see also machine intelligence

UNIVAC, 164
universal registration and
 identification system, 82
 see also behavioral monitoring;
  privacy; *and* surveillance

utopia, 81
 see also dystopia

war-gaming, 4, 119, 121
 see also computer applications,
  military
wayward robot, 41
 see also computers, conscious
  malevolence
work, xiii, xviii, 4, 82-83, 264
writing machines, 4
 see also computerized poetry

Selections Index: Names & Subjects

alienation, 273
Alvin, 104
anachronistic man, 285
AnCops, 125
Ariadnology, 182
artificial intelligence research, 209-211
Asimov, 244
Aub, Myron, 253
automated computer-design, 58, 257, 296
automated society
   authority, 140, 145-147
   computer-controlled economy, 145
   computer failure, 143
   freeloader fund-money, 139
   union featherbedding, 138
automatic writing, 59
autonomous technology, 72, 223

Balank, 283
Beast of Chorowait, 209
behavioral conditioning, 221-222
behavioral monitoring, 109, 160, 205-206
   see also surveillance
Billon, Professor, 131
Bohn 507 Computer, 131
Brant, Congressman, 253
Briskin, Jim, 137

Charles, Lieutenant, 275
Chementinski, 66

chess playing machine, 189-191
Chip, 98
Christina, 218
Chuck, 47
collapse of civilization, 178, 244
computer applications
   analysis of news, 17-23
   economic planning, 27
   government staffing, 27
   problem-solving, 38
   welfare, 171-177
   wish evaporation, 36
computer as
   god, 28-30, 80
   intellectual, 105
   person, 174
   servant, 116
computer capabilities, 8, 25
computer counselling, 175
computer domination, 297-300
computer government, 106
computer prediction, 74, 132
computer-run factories, 298
computer simulation, 106
computer verse, 57
computer war, 257
Computer West, 109
Computerdämmerung, 270
computerized elections (*see also* election forecasting)
   advantages, 150
   human factors, 155, 159
   patriotism, 162
   voter responsibility, 156

SELECTIONS INDEX: NAMES & SUBJECTS 343

computerized house, 235
computerized justice, 33
computerized political analysis, 148
computerized social control, 109
computers and
   consciousness, 8, 14
     see also consciousness *and*
       thinking machines
   creativity, 131
   emotional expression, 13
   free will, 8
   humor, 8, 11
   work, 271-273
     see also work
consciousness, 188
   see also computers and
control of research, 134
controlled agression, 124-130
crocodile of science, 133
cultural solidity, 273
Curver, Madison, 200
Cyfal, 283

Dark Brother, 293
Devlin, Dr., 200
Diablo, 17
discipline, 92-93
Dwar Ev, 80
Dwar Reyn, 80
dysfunctional technology, 224-225

EASCAC, 71
Eirkopf, Eblis, 66
election forecasting, 152, 268
   see also computerized elections
     and simulation of voting
     behavior
electromatic typewriters, 46
EPICAC XIV, 24
Epiktistes, 52
evolution of machines, 285, 295-296
expanding computer system, 219

fear of death, 207
Fischer, Maximillian, 138

Flamen, 17
freedom, 92-93, 207

games of chance, 105
George the booksweep, 67
goal-seeking behavior, 185
God's purpose, 48
Gondalug (the werewolf), 284
government computer
   natural resource, 159
   secrecy, 159, 160

Halyard, Ewing J., 24
Handley, Phil, 148
Hanley, George, 47
HOLMES FOUR, 7
home computer terminal, 173
homovivifying, 241
human degredation, 279-281
human problem, the, 298-300

identification and indexing, 87
   see also privacy; surveillance; *and*
     universal identifier
implanted monitoring device, 206
   see also behavioral monitoring
individual *vs.* society, 134
Industrial Revolution, 272
information implosion, 178-183
information storage and retrieval,
   178-183
innovation, 108
instant education, 241, 249, 251
   see also teaching machines
instinct, 245, 248, 250
intelligence testing, 94-97

Jaffe, Sam, 47
Joenes, 31, 209
Jordans, The, 94

Kapitza, Pyotr, 133
Katin, 270
Khashdrahr Miasma, 24
Khedron, 104

Ktistec machines, 55

lama, 45
Lambrick, Bob, 232
life
　as information, 104
　definition of, 187
Lynn, Jonathan, 26

machine breakdown, 176
machine classification, 102
machine future, 290
machine intelligence, 185-187
machine worship, 103
　see also synthetic religion
MALI (Manipulative Analysis and Logical Inference), 73
man against machine, 292-294
man as computer, 218
man as machine appendage, 275
man as servant, 268
man as slave, 275, 278
man-computer dialogue, 7-16, 52-62
man-computer hostility, 233
man-computer symbiosis, 296-297
man-machine comparisons, 25, 257, 268-270, 287
　see also evolution of machines
man-machine interaction, 219-220
　see also woman-computer liaison
Manny, 10
man's unpredictability, 269
Mark V Computer, 45
mechanical translation, 8
mental illness, causes, 273
Mike, 7
military computers, 253
Mill, John Stuart, 187
mind-force, 63
moral sense, 38
Mouse, 271
Moxon, 184
Mullen, Norman, 148
multidialed computing machine, 96
Multivac, 150

Murphy, Joe, 124

neural plugs, 271
neurally controlled machines, 272
NOCTIS (Non-Conceptual Thinking and Intuitional Synthesis), 74

Oliver, Matthew, 131
opinion polling, 201
Oracle at Sperry, 32
order and stability, 105

Papa Jan, 98
Perkins, Walter, 109
pocket computer, 254, 255
prevention of discovery, 133
privacy, 129, 237
　see also identification and idexing; surveillance; *and* universal identifier
progress, 259
psychosurgery, 215

Reconstruction and Reclamation Corps, 26
record-keeping, 87
research, basic *vs.* applied, 260
responsibility of scientists, 133, 261
revolt against the computer, 110

scanning device, 99
self-reproducing computers, 296-297
　see also automated computer-design
Senthree, 240
Servex, 218
Shackleton, Peter, 217
Shah of Bratpuhr, 24
Shuman Jehan, 253
simulation of (*see also* computer simulation)
　disorder, 124-130
　events, 20
　future, 132
　personality, 220-221

SELECTIONS INDEX: NAMES & SUBJECTS

simulation of *continued*
    voting behavior, 200-201
    see also election forecasting
Smirnov, Gregory, 52
social complexity, 93, 104
social engineering, 133
social stability, preconditions, 212-213
Software Man, 35
Spencer, Herbert, 187
Spielman, Max, 63
spoolpigeon show, 18
Stoker, Maurice, 66
Stranger, Richard, 217
super-hyper-adding machine, 280
supercalculator, 80
Supermind, 75
suppression of discovery, 133
surveillance, 88-89
    see also behavioral monitoring; identification and indexing; privacy; *and* universal identifier
synthetic religion, 213-214
    see also machine worship

teaching machines, 173-174, 218
    see also instant education
Tele-Pantographic Distorter, 59
teletypers, 137
thinking machines, 170, 184, 268
    see also computers and consciousness
thumb-marks, 90
trundler, 283

Unicephalon 40-D (homeostatic problem-solving structure), 137
UniComp, 98
universal identifer, 88
    see also identification and indexing; privacy; *and* surveillance
Utopia, 87, 93

Utopia *vs.* Liberalism, 88-89, 92-93

Wagner, Dr., 45
Warden, 9
Weider, General, 253
WESCAC, 63
Wieners, 170
woman-computer liaison, 239
    see also man-machine interaction
work, concept of, 272
    see also computers and work
work ethic, 276

Zero, 275